Problem Books in Mathematics

---

Edited by P. R. Halmos

Unsolved Problems in Intuitive Mathematics
Volume I

Richard K. Guy

# Unsolved Problems in Number Theory

With 17 Figures

Springer-Verlag
New York Heidelberg Berlin

Richard K. Guy
Department of Mathematics and Statistics
The University of Calgary
Canada T2N 1N4



AMS Classification (1980): 10-01

Library of Congress Cataloging in Publication Data
Guy, Richard K.
Unsolved problems in number theory.

(Unsolved problems in intuitive mathematics; v. 1)
(Problem books in mathematics)
Includes indexes.
1. Numbers, Theory of—Problems, exercises, etc.
I. Title. II. Series: Guy, Richard K. Unsolved problems in intuitive mathematics; v. 1. III. Series: Problem books in mathematics.
QA43.G88 vol. 1 [QA141] 510′.76s 81-14551
[512′.7′076] AACR2



Printed in the United States of America.

9 8 7 6 5 4 3 2 1

ISBN 0-387-90593-6 Springer-Verlag New York Heidelberg Berlin
ISBN 3-540-90593-6 Springer-Verlag Berlin Heidelberg New York

*For Erdős Pál*

*Among his several greatnesses are an ability to ask the right question and to ask it of the right person.*

# Preface

To many laymen, mathematicians appear to be problem solvers, people who do "hard sums". Even inside the profession we classify ourselves as either theorists or problem solvers. Mathematics is kept alive, much more than by the activities of either class, by the appearance of a succession of unsolved problems, both from within mathematics itself and from the increasing number of disciplines where it is applied. Mathematics often owes more to those who ask questions than to those who answer them. The solution of a problem may stifle interest in the area around it. But "Fermat's Last Theorem", because it is not yet a theorem, has generated a great deal of "good" mathematics, whether goodness is judged by beauty, by depth or by applicability.

To pose good unsolved problems is a difficult art. The balance between triviality and hopeless unsolvability is delicate. There are many simply stated problems which experts tell us are unlikely to be solved in the next generation. But we have seen the Four Color Conjecture settled, even if we don't live long enough to learn the status of the Riemann and Goldbach hypotheses, of twin primes or Mersenne primes, or of odd perfect numbers. On the other hand, "unsolved" problems may not be unsolved at all, or may be much more tractable than was at first thought.

Among the many contributions made by Hungarian mathematician Erdös Pál, not least is the steady flow of well-posed problems. As if these were not incentive enough, he offers rewards for the first solution of many of them, at the same time giving his estimate of their difficulty. He has made many payments, from \$1.00 to \$1000.00.

One purpose of this book is to provide beginning researchers, and others who are more mature, but isolated from adequate mathematical stimulus, with a supply of easily understood, if not easily solved, problems which they can consider in varying depth, and by making occasional partial progress, gradually acquire the interest, confidence and persistence that are essential to successful research.

But the book has a much wider purpose. It is important for students and teachers of mathematics at all levels to realize that although they are not yet capable of research and may have no hopes or ambitions in that direction, there are plenty of unsolved problems that are well within their comprehension, some of which will be solved in their lifetime. Many amateurs have been attracted to the subject and many successful researchers first gained their confidence by examining problems in euclidean geometry, in number theory, and more recently in combinatorics and graph theory, where it is possible to understand questions and even to formulate them and obtain original results without a deep prior theoretical knowledge.

The idea for the book goes back some twenty years, when I was impressed by the circulation of lists of problems by the late Leo Moser and co-author Hallard Croft, and by the articles of Erdös. Croft agreed to let me help him amplify his collection into a book, and Erdös has repeatedly encouraged and prodded us. After some time, the Number Theory chapter swelled into a volume of its own, part of a series which will contain a volume on Geometry, Convexity and Analysis, written by Hallard T. Croft, and one on Combinatorics, Graphs and Games by the present writer.

References, sometimes extensive bibliographies, are collected at the end of each problem or article surveying a group of problems, to save the reader from turning pages. In order not to lose the advantage of having all references collected in one alphabetical list, we give an Index of Authors, from which particular papers can easily be located provided the author is not too prolific. Entries in this index and in the General Index and Glossary of Symbols are to problem numbers instead of page numbers.

Many people have looked at parts of drafts, corresponded and made helpful comments. Some of these were personal friends who are no longer with us: Harold Davenport, Hans Heilbronn, Louis Mordell, Leo Moser, Theodor Motzkin, Alfred Rényi and Paul Turán. Others are H. L. Abbott, J. W. S. Cassels, J. H. Conway, P. Erdös, Martin Gardner, R. L. Graham, H. Halberstam, D. H. and Emma Lehmer, A. M. Odlyzko, Carl Pomerance, A. Schinzel, J. L. Selfridge, N. J. A. Sloane, E. G. Straus, H. P. F. Swinnerton-Dyer and Hugh Williams. A grant from the National (Science and Engineering) Research Council of Canada has facilitated contact with these and many others. The award of a Killam Resident Fellowship at The University of Calgary was especially helpful during the writing of a final draft. The technical typing was done by Karen McDermid, by Betty Teare and by Louise Guy, who also helped with proof-reading. The staff of Springer-Verlag in New York has been courteous, competent and helpful.

In spite of all this help, many errors remain, for which I assume reluctant responsibility. In any case, if the book is to serve its purpose it will start becoming out of date from the moment it appears; it has been becoming out of date ever since its writing began. I would be glad to hear from readers. There must be many solutions and references and problems which I don't know about. I hope that people will avail themselves of this clearing house. A few good researchers thrive by rediscovering results for themselves, but many of us are disappointed when we find that our discoveries have been anticipated.

Calgary 81:08:13 Richard K. Guy

# Glossary of Symbols

| | | |
|---|---|---|
| A.P. | arithmetic progression, $a, a+d, \ldots, a+kd, \ldots$ | A6, E10, E33 |
| $a_1 \equiv a_2 (\mathrm{mod}\ b)$ | $a_1$ congruent to $a_2$, modulo $b$; $a_1 - a_2$ divisible by $b$. | A3, A4, A12, A15, B21, ... |
| $A(x)$ | number of members of a sequence not exceeding $x$; e.g. number of amicable numbers not exceeding $x$ | B4, E1, E2, E4 |
| $c$ | a positive constant | A1, A3, A8, A12, B4, B11, ... |
| $d_n$ | difference between consecutive primes; $p_{n+1} - p_n$. | A8, A10, A11 |
| $d(n)$ | the number of (positive) divisors of $n$; $\sigma_0(n)$ | B, B2, B8, B12, B18, ... |
| $d \mid n$ | $d$ divides $n$; $n$ is a multiple of $d$; there is an integer $q$ such that $dq = n$. | B, B17, B32, B37, B44, C20, D2, E16 |
| $d \nmid n$ | $d$ does not divide $n$. | B, B2, B25, D2, E14, E16, ... |
| $e$ | base of natural logarithms; 2.718281828459045 ... | A8, B22, B39, D12, ... |
| $E_n$ | Euler numbers; coefficients in series for sec $x$. | B45 |

| | | |
|---|---|---|
| $\exp\{\ \}$ | exponential function. | A12, A19, B4, B36, B39, . . . |
| $F_n$ | Fermat numbers; $2^{2^n} + 1$. | A3, A12 |
| $f(x) \sim g(x)$ | $f(x)/g(x) \to 1$ as $x \to \infty$. | A3, C1, C17, E30, F26 |
| $f(x) = o(g(x))$ | $f(x)/g(x) \to 0$ as $x \to \infty$. | A1, A18, A19, B4, C11 |
| $f(x) = O(g(x))$ | there is a $c$ such that $\lvert f(x)\rvert < cg(x)$. | C8, C9, C10, C11, C12, . . . |
| $f(x) \ll g(x)$ | (same as above) | B4, B32, B40, C14, D11, E28, F4 |
| $i$ | square root of $-1$; $i^2 = -1$. | A16 |
| $\ln x$ | natural logarithm of $x$ | A1, A2, A3, A5, A8, A12, . . . |
| $(m, n)$ | g.c.d. (greatest common divisor) of $m$ and $n$; h.c.f. (highest common factor) of $m$ and $n$ | B3, B4, B5, B11, D2 |
| $[m, n]$ | l.c.m. (least common multiple) of $m$ and $n$. Also, the block of consecutive integers, $m, m+1, \ldots, n$. | B35, E2, F14; B24, B26, B32, C12, C16 |
| $M_n$ | Mersenne numbers, $2^n - 1$. | A3, B11, B38 |
| $n!$ | factorial $n$; $1 \times 2 \times 3 \times \cdots \times n$. | A2, B12, B14, B22, B23, B43, . . . |
| $\binom{n}{k}$ | $n$ choose $k$; the binomial coefficient $n!/k!(n-k)!$ | B31, B33, C10, D3 |
| $\left(\frac{p}{q}\right)$ | Legendre (or Jacobi) symbol. | see F5 (A1, A12, F7) |
| $p^a \parallel n$ | $p^a$ divides $n$, but $p^{a+1}$ does not divide $n$. | B, B8, B37, F16 |
| $p_n$ | the $n$th prime, $p_1 = 2$, $p_2 = 3, p_3 = 5, \ldots$ | A2, A5, A14, A17, E30 |
| $P(n)$ | largest prime factor of $n$. | B32, B46 |
| $\mathbb{Q}$ | the field of rational numbers | F7 |
| $r_k(n)$ | least number of numbers not exceeding $n$, which must contain a $k$-term A.P. | see E10 |

| | | |
|---|---|---|
| $s(n)$ | sum of aliquot parts (divisors of $n$ other than $n$) of $n$; $\sigma(n) - n$ | B, B1, B2, B8, B10, ... |
| $s^k(n)$ | $k$th iterate of $s(n)$ | B, B6, B7 |
| $s^*(n)$ | sum of unitary aliquot parts of $n$. | B8 |
| $W(k, l)$ | van der Waerden number. | see E10 |
| $\lfloor x \rfloor$ | floor of $x$; greatest integer not greater than $x$. | A1, A5, C7, C12, C15, ... |
| $\lceil x \rceil$ | ceiling of $x$; least integer not less than $x$. | B24 |
| $\mathbb{Z}_n$ | the ring of integers, 0, 1, 2, ..., $n - 1$ (modulo $n$) | E8 |
| $\gamma$ | Euler's constant; 0.577215664901532 ... | A8 |
| $\varepsilon$ | arbitrarily small positive constant. | A8, A18, A19, B4, B11, ... |
| $\pi$ | ratio of circumference of circle to diameter; 3.141592653589793 ... | F1, F17 |
| $\pi(x)$ | number of primes not exceeding $x$. | A17, E4 |
| $\pi(x; a, b)$ | number of primes not exceeding $x$ and congruent to $a$, modulo $b$. | A4 |
| $\prod$ | product | A1, A2, A3, A8, A15, ... |
| $\sigma(n)$ | sum of divisors of $n$; $\sigma_1(n)$. | B, B2, B5, B8, B9, ... |
| $\sigma_k(n)$ | sum of $k$th powers of divisors of $n$. | B, B12, B13, B14 |
| $\sigma^k(n)$ | $k$th iterate of $\sigma(n)$. | B9 |
| $\sigma^*(n)$ | sum of unitary divisors of $n$. | B8 |
| $\sum$ | sum | A5, A8, A12, B2, B14, ... |
| $\varphi(n)$ | Euler's totient function; number of numbers not exceeding $n$ and prime to $n$. | B8, B11, B36, B38, B39, ... |
| $\varphi^k(n)$ | $k$th iterate of $\varphi(n)$ | B41 |

| | | |
|---|---|---|
| $\omega$ | complex cube root of 1, $\omega^3 = 1, \omega \neq 1$, $\omega^2 + \omega + 1 = 0$. | A16 |
| $\omega(n)$ | number of distinct prime factors of $n$. | B2, B8 |
| $\Omega(n)$ | number of prime factors of $n$, counting repetitions. | B8 |
| ¿ · · · ? | conjectural or hypothetical statement. | A1, A9, B37, C6, E10, E28, F2, F18 |

# Contents

# Introduction

Number theory has fascinated both the amateur and the professional for a longer time than any other branch of mathematics; so that much of it is now of considerable technical difficulty. However, there are more unsolved problems than ever before, and though many of these are unlikely to be solved in the next generation, this probably won't deter people from trying. They are so numerous that they have already filled more than one volume so that the present book is just a personal sample.

Erdös recalls that Landau, at the International Congress in Cambridge in 1912, gave a talk about primes and mentioned four problems (see A1, A7, C1 below) which were unattackable in the present state of science, and says that in 1980 they still are.

Here are some good sources of problems in number theory.


P. Erdös, Some unsolved problems, *Michigan Math. J.* **4** (1957) 291–300.

P. Erdös, On unsolved problems, *Publ. Math. Inst. Hungar. Acad. Sci.* **6** (1961) 221–254.

P. Erdös, *Quelques Problèmes de la Théorie des Nombres*, Monographies de l'Enseignment Math. #6, Geneva, 1963, 81–135.

P. Erdös, Extremal problems in number theory, *Proc. Symp. Pure Math.* **8**, Amer. Math. Soc., Providence, 1965, 181–189.

P. Erdös, Some recent advances and current problems in number theory, in *Lectures on Modern Mathematics* **3**, Wiley, New York, 1965, 196–244.

P. Erdös, Résultats et problèmes en théorie des nombres, *Seminar Delange-Pisot-Poitou* **24**, 1972–73.

P. Erdös, Problems and results in combinatorial number theory, in *A Survey of Combinatorial Theory*, North-Holland, 1973, 117–138.

P. Erdös, *Problems and Results in Combinatorial Number Theory*, Bordeaux, 1974.

Paul Erdös, Problems and results in combinatorial number theory III, *Springer Lecture Notes in Math.* **626** (1977) 43–72; *MR* **57** #12442.

P. Erdös, Combinatorial problems in geometry and number theory, *Amer. Math. Soc. Proc. Sympos. Pure Math.* **34** (1979) 149–162.

Paul Erdös, A survey of problems in combinatorial number theory, in *Combinatorial Mathematics, Optimal Designs and their Applications* (Proc. Symp. Colo. State Univ. 1978) Ann. Discrete Math. **6** (1980) 89–115.
Paul Erdös, Problems and results in number theory and graph theory, *Congressus Numerantium* XXVII (Proc. 9th Manitoba Conf. Num. Math. Comput. 1979) Utilitas Math., Winnipeg, 1980, 3–21.
P. Erdös and R. L. Graham, *Old and New Problems and Results in Combinatorial Number Theory*, Monographies de l'Enseignment Math. No. 28, Geneva, 1980.
Pál Erdös and András Sárközy, Some solved and unsolved problems in combinatorial number theory, *Math. Slovaca*, **28** (1978) 407–421; *MR* **80i**:10001.
H. Fast and S. Świerczkowski, *The New Scottish Book*, Wrocław, 1946–1958.
Heini Halberstam, Some unsolved problems in higher arithmetic, in Ronald Duncan and Miranda Weston-Smith (eds.) *The Encyclopaedia of Ignorance*, Pergamon, Oxford and New York, 1977, 191–203.
*Proceedings of Number Theory Conference*, Univ. of Colorado, Boulder, 1963.
*Report of Institute in the Theory of Numbers*, Univ. of Colorado, Boulder, 1959.
Daniel Shanks, *Solved and Unsolved Problems in Number Theory*, Chelsea, New York, 2nd ed. 1978; *MR* **80e**:10003.
W. Sierpiński, *A Selection of Problems in the Theory of Numbers*, Pergamon, 1964.
S. Ulam, *A Collection of Mathematical Problems*, Interscience, New York, 1960.

Throughout this volume, "number" means natural number, $c$ is an absolute positive constant, not necessarily the same each time it appears, and $\varepsilon$ is an arbitrarily small positive constant. We use Donald Knuth's "floor" ($\lfloor\ \rfloor$) and "ceiling ($\lceil\ \rceil$) symbols for "the greatest integer not greater than" and "the least integer not less than."

The notation $f(x) = O(g(x))$ and $f(x) \ll g(x)$ mean that there are constants $c_1$, $c_2$ such that $c_1 g(x) < f(x) < c_2 g(x)$ for all sufficiently large $x$; while $f(x) \sim g(x)$ means that $f/g \to 1$, and $f(x) = o(g(x))$ means that $f/g \to 0$, as $x \to \infty$.

The book has been partitioned, somewhat arbitrarily at times, into six sections:

A. Prime numbers
B. Divisiblity
C. Additive number theory
D. Diophantine equations
E. Sequences of integers
F. None of the above.

# A. Prime Numbers

We can partition the positive integers into three classes:

the unit, 1
the primes, 2, 3, 5, 7, 11, 13, 17, 19, 23, 29, 31, 37, . . .
the composite numbers, 4, 6, 8, 9, 10, . . .

A number greater than 1 is **prime** if its only positive divisors are 1 and itself; otherwise it's **composite**. Primes have interested mathematicians at least since Euclid, who showed that there were infinitely many.

Denote the $n$th prime by $p_n$, e.g., $p_1 = 2, p_2 = 3, p_{99} = 523$; and the number of primes not greater than $x$ by $\pi(x)$, e.g., $\pi(2) = 1$, $\pi(3\frac{1}{2}) = 2$, $\pi(1000) = 168$. The greatest common divisor (g.c.d.) of $m$ and $n$ is denoted by $(m, n)$. If $(m, n) = 1$, we say that $m$ and $n$ are **coprime**; for example, $(14, 15) = 1$.

Dirichlet's theorem tells us that there are infinitely many primes in any arithmetic progression,

$$a, \ a + b, \ a + 2b, \ a + 3b, \ldots$$

provided $(a, b) = 1$. An article, giving a survey of problems about primes and a number of further references, is

A. Schinzel and W. Sierpiński, Sur certains hypothèses concernant les nombres premiers, *Acta Arith.* **4** (1958) 185–208 (erratum **5** (1959) 259); *MR* **21** #4936.

Table 7 (D27) can be used as a table of primes <1000.

The general problem of determining whether a large number is prime or composite, and in the latter case of determining its factors, has fascinated number theorists down the ages. With the advent of high speed computers, considerable advances have been made, and a special stimulus has recently been provided by the application to cryptanalysis. Some other references appear after Problem A3.

Leonard Adleman and Frank Thomson Leighton, An $O(n^{1/10.89})$ primality testing algorithm, *Math. Comput.* **36** (1981) 261–266.
Leonard M. Adleman, Carl Pomerance and Robert S. Rumely, On distinguishing prime numbers from composite numbers (to appear)
R. P. Brent, An improved Monte Carlo factorization algorithm, *BIT*, **20** (1980), 176–184.
John D. Dixon, Asymptotically fast factorization of integers, *Math. Comput.* **36** (1981) 255–260.
Richard K. Guy, How to factor a number, *Congressus Numerantium XVI* Proc. 5th Manitoba Conf. Numer. Math., Winnipeg, 1975, 49–89.
H. W. Lenstra, Primality testing, *Studieweek Getaltheorie en Computers*, Stichting Mathematisch Centrum, Amsterdam, 1980, 41–60.
G. L. Miller, Riemann's hypothesis and tests for primality, *J. Comput. System Sci.*, **13** (1976) 300–317.
J. M. Pollard, Theorems on factorization and primality testing, *Proc. Cambridge Philos. Soc.* **76** (1974) 521–528.
J. M. Pollard, A Monte Carlo method for factorization, *BIT* **15** (1975) 331–334; *MR* **50** #6992.
R. Rivest, A. Shamir and L. Adleman, A method for obtaining digital signatures and public key cryptosystems, *Communications A.C.M.*, Feb. 1978.
R. Solovay and V. Strassen, A fast Monte-Carlo test for primality, *SIAM J. Comput.* **6** (1977) 84–85; erratum **7** (1978) 118; *MR* **57** #5885.
H. C. Williams, Primality testing on a computer, *Ars Combin.* **5** (1978) 127–185. *MR* **80d**:10002.
H. C. Williams and R. Holte, Some observations on primality testing, *Math. Comput.* **32** (1978) 905–917; *MR* **57** #16184.
H. C. Williams and J. S. Judd, Some algorithms for prime testing using generalized Lehmer functions, *Math. Comput.* **30** (1976) 867–886.

**A1.** Are there infinitely many primes of the form $a^2 + 1$? Probably so, and in fact Hardy and Littlewood (their Conjecture E) guessed that the number, $P(n)$, of such primes less than $n$, was asymptotic to $c\sqrt{n}/\ln n$,

$$¿ \qquad P(n) \sim c\sqrt{n}/\ln n \qquad ?$$

i.e., that the ratio of $P(n)$ to $\sqrt{n}/\ln n$ tends to $c$ as $n$ tends to infinity. The constant $c$ is

$$c = \prod\left\{1 - \frac{\left(\frac{-1}{p}\right)}{p-1}\right\} = \prod\left\{1 - \frac{(-1)^{(p-1)/2}}{p-1}\right\} \approx 1.3727$$

where $\left(\frac{-1}{p}\right)$ is the Legendre symbol (see F5) and the product is taken over all odd primes. They make similar conjectures, differing only in the value of $c$, for the number of primes represented by more general quadratic expressions. But we don't know of any integer polynomial, of degree greater than one, for which it has been proved that it takes an infinity of prime values. Is there even one prime $a^2 + b$ for each $b > 0$?

Iwaniec has shown that there are infinitely many $n$ for which $n^2 + 1$ is the product of at most two primes, and his result extends to other irreducible quadratics.

Ulam and others noticed that the patterns formed by the prime numbers when the sequence of numbers is written in a "square spiral" seems to favor diagonals which correspond to certain "prime-rich" quadratic polynomials. For example, the main diagonal of Figure 1 corresponds to Euler's famous formula $n^2 + n + 41$.

| **421** | 420 | **419** | 418 | 417 | 416 | 415 | 414 | 413 | 412 | 411 | 410 | **409** | 408 | 407 | 406 | 405 | 404 | 403 | 402 |
|---|---|---|---|---|---|---|---|---|---|---|---|---|---|---|---|---|---|---|---|
| 422 | **347** | 346 | 345 | 344 | 343 | 342 | 341 | 340 | 339 | 338 | **337** | 336 | 335 | 334 | 333 | 332 | **331** | 330 | **401** |
| 423 | 348 | **281** | 280 | 279 | 278 | **277** | 276 | 275 | 274 | 273 | 272 | **271** | 270 | **269** | 268 | 267 | 266 | 329 | 400 |
| 424 | **349** | 282 | **223** | 222 | 221 | 220 | 219 | 218 | 217 | 216 | 215 | 214 | 213 | 212 | **211** | 210 | 265 | 328 | 399 |
| 425 | 350 | **283** | 224 | **173** | 172 | 171 | 170 | 169 | 168 | **167** | 166 | 165 | 164 | **163** | 162 | 209 | 264 | 327 | 398 |
| 426 | 351 | 284 | 225 | 174 | **131** | 130 | 129 | 128 | **127** | 126 | 125 | 124 | 123 | 122 | 161 | 208 | **263** | 326 | **397** |
| 427 | 352 | 285 | 226 | 175 | 132 | **97** | 96 | 95 | 94 | 93 | 92 | 91 | 90 | 121 | 160 | 207 | 262 | 325 | 396 |
| 428 | **353** | 286 | **227** | 176 | 133 | 98 | **71** | 70 | 69 | 68 | **67** | 66 | **89** | 120 | 159 | 206 | 261 | 324 | 395 |
| 429 | 354 | 287 | 228 | 177 | 134 | 99 | 72 | **53** | 52 | 51 | 50 | 65 | 88 | 119 | 158 | 205 | 260 | 323 | 394 |
| 430 | 355 | 288 | **229** | 178 | 135 | 100 | **73** | 54 | **43** | 42 | 49 | 64 | 87 | 118 | **157** | 204 | 259 | 322 | 393 |
| **431** | 356 | 289 | 230 | **179** | 136 | **101** | 74 | 55 | 44 | **41** | 48 | 63 | 86 | 117 | 156 | 203 | 258 | 321 | 392 |
| 432 | 357 | 290 | 231 | 180 | **137** | 102 | 75 | 56 | 45 | 46 | **47** | 62 | 85 | 116 | 155 | 202 | **257** | 320 | 391 |
| **433** | 358 | 291 | 232 | **181** | 138 | **103** | 76 | 57 | 58 | **59** | 60 | **61** | 84 | 115 | 154 | 201 | 256 | 319 | 390 |
| 434 | **359** | 292 | **233** | 182 | **139** | 104 | 77 | 78 | **79** | 80 | 81 | 82 | **83** | 114 | 153 | 200 | 255 | 318 | **389** |
| 435 | 360 | **293** | 234 | 183 | 140 | 105 | 106 | **107** | 108 | **109** | 110 | 111 | 112 | **113** | 152 | **199** | 254 | **317** | 388 |
| 436 | 361 | 294 | 235 | 184 | 141 | 142 | 143 | 144 | 145 | 146 | 147 | 148 | **49** | 150 | **151** | 198 | 253 | 316 | 387 |
| 437 | 362 | 295 | 236 | 185 | 186 | 187 | 188 | 189 | 190 | **191** | 192 | **193** | 194 | 195 | 196 | **197** | 252 | 315 | 386 |
| 438 | 363 | 296 | 237 | 238 | **239** | 240 | **241** | 242 | 243 | 244 | 245 | 246 | 247 | 248 | 249 | 250 | **251** | 314 | 385 |
| **439** | 364 | 297 | 298 | 299 | 300 | 301 | 302 | 303 | 304 | 305 | 306 | **307** | 308 | 309 | 310 | **311** | 312 | **313** | 384 |
| 440 | 365 | 366 | **367** | 368 | 369 | 370 | 371 | 372 | **373** | 374 | 375 | 376 | 377 | 378 | **379** | 380 | 381 | 382 | **383** |

Figure 1. Primes (in **bold**) Form Diagonal Patterns.

The only result for expressions (*not* polynomials!) of degree greater than 1 is due to Pyateckii-Šapiro, who proved that the number of primes of the form $\lfloor n^c \rfloor$ in the range $1 < n < x$ is $(1 + o(1))x/(1 + c)\ln x$ if $1 \leq c \leq 12/11$.

Martin Gardner, The remarkable lore of prime numbers, *Scientific Amer.* **210** #3 (Mar. 1964) 120–128.

G. H. Hardy and J. E. Littlewood, Some problems of 'partitio numerorum' III: on the expression of a number as a sum of primes, *Acta Math.* **44** (1922) 1–70.

Henryk Iwaniec, Almost-primes represented by quadratic polynomials, *Invent. Math.* **47** (1978) 171–188; *MR* **58** #5553.

Carl Pomerance, A note on the least prime in an arithmetic progression, *J. Number Theory* **12** (1980) 218–223.

I. I. Pyateckii-Šapiro, On the distribution in sequences of the form $[f(n)]$, *Mat. Sbornik N.S.* **33** (1953) 559–566; *MR* **15**, 507.

**A2.** Are there infinitely many primes of the form $n! + 1$? The only values of $n \leq 230$ which give primes are 1, 2, 3, 11, 27, 37, 41, 73, 77, 116, and 154. It is not known if $n! - 1$ or

$$X = 1 + \prod_{i=1}^{k} p_i$$

is prime infinitely often. The only values of $p_k \leq 1031$ for which $X$ is prime are $p_k = 2, 3, 5, 7, 11, 31, 379, 1019$, and 1021.

Let $q$ be the least prime greater than $X$. Then R. F. Fortune conjectures that $q - X + 1$ is prime for all $k$. It is clear that it is not divisible by the first $k$ primes, and Selfridge observes that the truth of the conjecture would follow from one of Schinzel, that for $x > 8$ there is always a prime between $x$ and $x + (\ln x)^2$. The first few fortunate primes are 3, 5, 7, 13, 23, 17, 19, 23, 37, 61, 67, 61, 71, 47, 107, 59, 61, 109, 89, 103, 79, . . . . The answers to the questions are probably "yes," but it does not seem conceivable that such conjectures will come within reach either of computers or of analytical tools in the foreseeable future.

More hopeful, but still difficult, is the following conjecture of Erdös and Stewart: are $1! + 1 = 2$, $2! + 1 = 3$, $3! + 1 = 7$, $4! + 1 = 5^2$, $5! + 1 = 11^2$ the only cases where $n! + 1 = p_k^a p_{k+1}^b$ and $p_{k-1} \leq n < p_k$? [Note that $(a, b) = (1, 0)$, $(1, 0)$, $(0, 1)$, $(2, 0)$, and $(0, 2)$ in these five cases.]

Erdös also asks if there are infinitely many primes $p$ for which $p - k!$ is composite for each $k$ such that $1 \leq k! < p$; for example, $p = 101$ and $p = 211$. He suggests that it may be easier to show that there are infinitely many integers $n$ $(l! < n \leq (l + 1)!)$ all of whose prime factors are greater than $l$, and for which all the numbers $n - k!$ $(1 \leq k \leq l)$ are composite.

David Silverman noticed that the product

$$\prod_{i=1}^{m} \frac{p_i + 1}{p_i - 1}$$

is an integer for $m = 1, 2, 3, 4$ and 8 and asked if it ever is again.

I. O. Angell and H. J. Godwin, Some factorizations of $10^n \pm 1$, *Math. Comput.* **28** (1974) 307–308.

Alan Borning, Some results for $k! \pm 1$ and $2 \cdot 3 \cdot 5 \cdots p \pm 1$, *Math. Comput.* **26** (1972) 567–570.

Martin Gardner, Mathematical Games, *Sci. Amer.* **243** #6 (Dec. 1980) 18–28.

Solomon W. Golomb, On Fortune's conjecture, *Math. Mag.* (to appear)

S. Kravitz and D. E. Penney, An extension of Trigg's table, *Math. Mag.* **48** (1975) 92–96.

Mark Templer, On the primality of $k! + 1$ and $2 * 3 * 5 * \cdots * p + 1$, *Math. Comput.* **34** (1980) 303–304.

**A3.** Primes of special form have been of perennial interest, especially the **Mersenne primes** $2^p - 1$ ($p$ is necessarily prime, but that is *not* sufficient! $2^{11} - 1 = 2047 = 23 \times 89$) in connexion with perfect numbers (see B1) and **repunits**, $(10^p - 1)/9$.

The powerful Lucas-Lehmer test, in conjunction with successive generations of computers, and more sophisticated techniques in using them, continues to add to the list of primes for which $2^p - 1$ is also prime:

$$2, 3, 5, 7, 13, 17, 19, 31, 61, 89, 107, 127, 521, 607, 1279, 2203,$$
$$2281, 3217, 4253, 4423, 9689, 9941, 11213, 19937, 21701, 23209, 44497, \ldots .$$

Their number is undoubtedly infinite, but proof is again hopelessly beyond reach. Suppose $M(x)$ is the number of primes $p \le x$ for which $2^p - 1$ is prime. Find a convincing heuristic argument for the size of $M(x)$. Gillies gave one suggesting that $M(x) \sim c \ln x$, but some people do not believe this. Pomerance has an argument for $M(x) \sim c(\ln \ln x)^2$ but he says this doesn't agree with the facts.

D. H. Lehmer puts $S_1 = 4$, $S_{k+1} = S_k^2 - 2$, supposes that $2^p - 1$ is a Mersenne prime, notes that $S_{p-2} \equiv 2^{(p+1)/2}$ or $-2^{(p+1)/2} \pmod{2^p - 1}$ and asks: which?

Selfridge conjectures that if $n$ is a prime of the form $2^k \pm 1$ or $2^{2k} \pm 3$, then $2^n - 1$ and $(2^n + 1)/3$ are either both prime or neither of them are. Moreover if both are prime, then $n$ is of one of those forms. Is this an example of "the strong law of small numbers"?

If $p$ is a prime, is $2^p - 1$ always **squarefree** (does it never contain a repeated factor)? This seems to be another unanswerable question. It is safe to conjecture that the answer is "No!" This *could* be settled by computer if you were lucky. As D. H. Lehmer has said about various factorization methods, "Happiness is just around the corner." Selfridge puts the computational difficulties in perspective by proposing the problem: find fifty more numbers like 1093 and 3511. (Fermat's theorem tells us that if $p$ is prime, then $p$ divides $2^p - 2$; the primes 1093 and 3511 are the only ones less than $3 \times 10^9$ for which $p^2$ divides $2^p - 2$.)

The corresponding primes for $(10^p - 1)/9$ are 2, 19, 23, 317, 1031, the last two of which were found by Hugh Williams quite recently, subject to final tests being completed in the last case. Repunits $>1$ are known never to be squares. Are they ever cubes? When are they squarefree?

The **Fermat numbers**, $F_n = 2^{2^n} + 1$, are also of continuing interest; they are prime for $0 \le n \le 4$ and composite for $5 \le n \le 19$ and for many larger values of $n$. Hardy and Wright give a heuristic argument which suggests that only a finite number of them are prime. Selfridge would like to see this strengthened to support the conjecture that all the rest are composite.

Because of their special interest as potential factors of Fermat numbers, and because proofs of their primality are comparatively easy, numbers of the form $k \cdot 2^n + 1$ have received special attention, at least for small values of $k$. For example, Hugh Williams found that if $k = 5$, then $n = 3313$, 4687 and 5947 give primes, the first of which divides $F_{3310}$, and Richard Brent has shown that $p = 1238926361552897$ divides $F_8$. See also B21.

We are very unlikely to know for sure that the **Fibonacci sequence**

1, 1, **2**, **3**, **5**, 8, **13**, 21, 34, 55, **89**, 144, **233**, 377, 610, 987, **1597**, . . .

where $u_1 = u_2 = 1, u_{n+1} = u_n + u_{n-1}$ $(n \geq 2)$ contains infinitely many primes. Similarly for the related **Lucas sequence**

1, **3**, 4, **7**, **11**, 18, **29**, **47**, 76, 123, **199**, 322, **521**, 843, 1364, . . .

and most other Lucas-Lehmer sequences (with $(u_1, u_2) = 1$) defined by second order recurrence relations. However, Graham has shown that the sequence with

$$u_1 = 1786\ 772701\ 928802\ 632268\ 715130\ 455793$$
$$u_2 = 1059\ 683225\ 053915\ 111058\ 165141\ 686995$$

contains no primes at all!

Raphael Robinson considers the Lucas sequence $u_0 = 0$, $u_1 = 1$, $u_{n+1} = 2u_n + u_{n-1}$ $(n \geq 1)$ and defines the **primitive part**, $L_n$, by

$$u_n = \prod_{d|n} L_n$$

He notes that $L_7 = 13^2$ and $L_{30} = 31^2$ and asks if there is any $n > 30$ for which $L_n$ is a square.


George E. Andrews, Some formulae for the Fibonacci sequence with generalizations, *Fibonacci Quart.* **7** (1969) 113–130; *MR* **39** #4088.

R. C. Archibald, *Scripta Math.* **3** (1935) 117.

Robert Baillie, New primes of the form $k \cdot 2^n + 1$, *Math. Comput.* **33** (1979) 1333–1336; *MR* **80h**: 10009.

Richard P. Brent, Factorization of the eighth Fermat number, *Abstracts Amer. Math. Soc.* **1** (1980) 565.

Richard P. Brent and J. M. Pollard, Factorization of the eighth Fermat number, *Math. Comput.* **36** (1981)

John Brillhart, D. H. Lehmer and J. L. Selfridge, New primality criteria and factorizations of $2^m \pm 1$, *Math. Comput.* **29** (1975) 620–627; *MR* **52** #5546.

John Brillhart, D. H. Lehmer, J. L. Selfridge, Bryant Tuckerman and S. S. Wagstaff, Factorizations of $b^n - 1$ and $b^n + 1$, $b < 13$, *Amer. Math. Soc.*, Providence, 1981, (to appear)

J. Brillhart, J. Tonascia and P. Weinberger, On the Fermat quotient, in A. O. L. Atkin and B. J. Birch (eds.) *Computers in Number Theory*, Academic Press, London, 1971, pp. 213–222.

Martin Gardner, Mathematical games: The strong law of small numbers, *Sci. Amer.* **243** #6 (Dec. 1980), 18–28.

Donald B. Gillies, Three new Mersenne primes and a statistical theory, *Math. Comput.* **18** (1964) 93–97.

Gary B. Gostin, A factor of $F_{17}$, *Math. Comput.* **35** (1980) 975–976.

Gary B. Gostin and Philip B. McLaughlin, Five new factors of Fermat numbers, *Math. Comput.*, (to appear)

R. L. Graham, A. Fibonacci-like sequence of composite numbers, *Math. Mag.* **37** (1964) 322–324.

John C. Hallyburton and John Brillhart, Two new factors of Fermat numbers, *Math. Comput.* **29** (1975) 109–112; *MR* **51** #5460. Corrigendum **30** (1976) 198; *MR* **52** #13599.
M Kraitchik, *Sphinx*, 1931, 31.
D. H. Lehmer, *Sphinx*, 1931, 32, 164.
E. Lucas, Théorie des fonctions numériques simplements périodiques, *Amer. J. Math.* **1** (1878) 184–240, 289–321 (esp. p. 316)
G. Matthew and H. C. Williams, Some new primes of the form $k \cdot 2^n + 1$, *Math. Comput.* **31** (1977) 797–798; *MR* **55** #12605.
Michael A. Morrison and John Brillhart, A method of factoring and the factorization of $F_7$, *Math. Comput.* **29** (1975) 183–205.
Curt Noll and Laura Nickel, The 25th and 26th Mersenne primes, *Math. Comput.* **35** (1980) 1387–1390.
Rudolf Ondrejka, Primes with 100 or more digits, *J. Recreational Math.* **2** (1969) 42–44; Addenda, **3** (1970) 161–162; More on large primes, **11** (1979) 112–113; *MR* **80g**:10012.
R. E. Powers, *Amer. Math. Monthly*, **18** (1911) 195–197; *Proc. London Math. Soc.* (2) **13** (1919) 39.
Raphael M. Robinson, A report on primes of the form $k \cdot 2^n + 1$ and on factors of Fermat numbers, *Proc. Amer. Math. Soc.* **9** (1958) 673–681; *MR* **20** #3097.
D. E. Shippee, Four new factors of Fermat numbers, *Math. Comput.* **32** (1978) 941.
David Slowinski, Searching for the 27th Mersenne prime, *J. Recreational Math.* **11** (1978–79) 258–261; *MR* **80g**:10013.
C. L. Stewart, The greatest prime factor of $A^n - B^n$, *Acta Arith.* **26** (1975) 427–433.
C. L. Stewart, Divisor properties of arithmetical sequences, PhD. thesis, Cambridge, 1976.
Bryant Tuckerman, The 24th Mersenne prime, *Proc. Nat. Acad. Sci. U.S.A.* **68** (1971) 2319–2320.
D. D. Wall, Fibonacci series modulo *m*, *Amer. Math. Monthly* **67** (1960) 525–532; *MR* **22** #10945.
H. C. Williams, Some primes with interesting digit patterns, *Math. Comput.* **32** (1978) 1306–1310; *MR* **58** #484; Zbl. 388.10007.
H. C. Williams and E. Seah, Some primes of the form $(a^n - 1)/(a - 1)$, *Math. Comput.* **33** (1979) 1337–1342; *MR* **80g**:10014.
Samuel Yates, The mystique of repunits, *Math. Mag.* **51** (1978) 22–28; *MR* **80f**:10008.

**A4.** A number $a$ is said to be **congruent** to $c$, **modulo** a positive number $b$, written $a \equiv c \pmod{b}$, if $b$ is a divisor of $a - c$. S. Chowla conjectured that if $(a, b) = 1$, then there are infinitely many pairs of consecutive primes, $p_n$, $p_{n+1}$ such that $p_n \equiv p_{n+1} \equiv a \pmod{b}$. The case $b = 4$, $a = 1$ follows from a theorem of Littlewood. Bounds between which such consecutive primes occur have been given in this case, and for $b = 4$, $a = 3$ by Knapowski and Turán. Turán observed that it would be of interest (in connexion with the Riemann hypothesis, for example) to discover long sequences of consecutive primes $\equiv 1 \pmod{4}$. Den Haan found the 9 primes

11593, 11597, 11617, 11621, 11633, 11657, 11677, 11681, 11689.

The next longer sequence contains the 11 primes

766261, 766273, 766277, 766301, 766313, 766321, 766333, 766357, 766361, 766369, 766373.

Turán was particularly interested in the **prime number race**. Let $\pi(n; a, b)$ be the number of primes $p < n$, $p \equiv a \pmod{b}$. Is it true that for every $a$ and $b$ with $(a, b) = 1$, there are infinitely many values of $n$ for which

$$\pi(n;\, a, b) > \pi(n;\, a_1, b)$$

for every $a_1 \not\equiv a \pmod{b}$? Knapowski and Turán settled special cases, but the general problem is wide open.

Chebyshev noted that $\pi(n; 1, 3) < \pi(n; 2, 3)$ and $\pi(n; 1, 4) < \pi(n; 3, 4)$ for small values of $n$. Leech, and independently Shanks and Wrench, discovered that the second inequality is reversed for $n = 26861$ and Bays and Hudson that the first is reversed for two sets, each of more than 150 million integers, between $n = 608981813029$ and $n = 610968213796$.

Carter Bays and Richard H. Hudson, The appearance of tens of billions of integers $x$ with $\pi_{24,13}(x) < \pi_{24,1}(x)$ in the vicinity of $10^{12}$, *J. reine angew. Math.* **299/300** (1978) 234–237; *MR* **57** #12418.

Carter Bays and Richard H. Hudson, Details of the first region of integers $x$ with $\pi_{3,2}(x) < \pi_{3,1}(x)$, *Math. Comput.* **32** (1978) 571–576.

Carter Bays and Richard H. Hudson, Numerical and graphical description of all axis crossing regions for the moduli 4 and 8 which occur before $10^{12}$, *Internat. J. Math. Sci.* **2** (1979) 111–119; *MR* **80h**:10003.

Richard H. Hudson, A common combinatorial principle underlies Riemann's formula, the Chebyshev phenomenon, and other subtle effects in comparative prime number theory I, *J. reine angew. Math.* **313** (1980) 133–150.

S. Knapowski and P. Turán, Über einige Fragen der vergleichenden Primzahltheorie, *Number Theory and Analysis*, Plenum Press, New York, 1969, 157–171.

S. Knapowski and P. Turán, On prime numbers $\equiv 1$ resp. 3 (mod 4), *Number Theory and Algebra*, Academic Press, N.Y. 1977, pp. 157–165; *MR* **57** #5926.

John Leech, Note on the distribution of prime numbers, *J. London Math.* Soc. **32** (1957) 56–58.

Daniel Shanks, Quadratic residues and the distribution of primes, *Math. Tables Aids Comput.* **13** (1959) 272–284.

**A5.** How long can an arithmetic progression be which consists only of primes? Table 1 shows progressions of $n$ primes, $a, a + d, \ldots, a + (n - 1)d$, discovered by V. A. Golubev, E. Karst, S. C. Root, W. N. Seredinskii, and S. Weintraub. Of course, the common difference must have every prime $\leq n$ as a divisor (unless $n = a$). It is conjectured that $n$ can be as large as you like. This would follow if it were possible to improve Szemerédi's theorem (see E10).

More generally, Erdös conjectures that if $\{a_i\}$ is any infinite sequence of integers for which $\sum 1/a_i$ is divergent, then the sequence contains arbitrarily long arithmetic progressions. He offers \$3000 for a proof or disproof of this conjecture.

Sierpiński defines $g(x)$ to be the maximum number of terms in a progression of primes not greater than $x$. The least $x$, $l(x)$, for which $g(x)$ takes the values

$$g(x) = 0 \quad 1 \quad 2 \quad 3 \quad 4 \quad 5 \quad 6 \quad 7 \quad 8 \quad 9 \quad 10 \ldots$$

is

$$l(x) = 1 \quad 2 \quad 3 \quad 7 \quad 23 \quad 29 \quad 157 \quad 1307 \quad 1669 \quad 1879 \quad 089 \ldots .$$

Table 1. Long Arithmetic Progressions of Primes.

| $n$ | $d$ | $a$ | $a+(n-1)d$ | discovery |
|---|---|---|---|---|
| 12 | 11550 | 166601 | 293651 | K, 1967 |
| 12 | 13860 | 110437 | 262897 | K, 1967 |
| 12 | 13860 | 152947 | 305407 | K, 1967 |
| 12 | 30030 | 23143 | 353473 | G, 1958 |
| 12 | 30030 | 1498141 | 1829471 | K, 1968 |
| 12 | 30030 | 188677831 | 189008161 | R, 1969 |
| 12 | 30030 | 805344823 | 805675153 | R, 1969 |
| 12 | 90090 | 409027 | 1400017 | S, 1966 |
| 12 | 90090 | 802951 | 1793941 | K, 1969 |
| 12 | 90090 | 862397 | 1853387 | K, 1969 |
| 13 | 60060 | 4943 | 725663 | |
| 13 | 510510 | 766439 | 6892559 | S, 1965 |
| 14 | 2462460 | 46883579 | 78895559 | |
| 16 | 9699690 | 53297929 | 198793279 | |
| 16 | 223092870 | 2236133941 | 5582526991 | R, 1969 |
| 17 | 87297210 | 3430751869 | 4827507229 | W, 1977 |

Pomerance produces the "prime number graph" by plotting the points $(n, p_n)$ and shows that for every $k$ we can find $k$ primes whose points are collinear.

Grosswald has shown that there are long arithmetic progressions consisting only of **almost primes**, in the following sense. There are infinitely many arithmetic progressions of $k$ terms, each term being the product of at most $r$ primes, where

$$r \leq \lfloor k \ln k + 0.892k + 1 \rfloor.$$

P. Erdös and P. Turán, On certain sequences of integers, *J. London Math. Soc.* **11** (1936) 261–264.

J. Gerver, The sum of the reciprocals of a set of integers with no arithmetic progression of $k$ terms, *Proc. Amer. Math. Soc.* **62** (1977) 211–214.

Joseph L. Gerver and L. Thomas Ramsey, Sets of integers with no long arithmetic progressions generated by the greedy algorithm, *Math. Comput.* **33** (1979) 1353–1359.

V. A. Golubev, Faktorisation der Zahlen der Form $x^3 \pm 4x^2 + 3x \pm 1$, *Anz. Oesterreich. Akad. Wiss. Math.-Naturwiss. Kl.* 1969, 184–191 (see also 191–194; 297–301; 1970, 106–112; 1972, 19–20, 178–179).

Emil Grosswald, Long arithmetic progressions that consist only of primes and almost primes, *Notices Amer. Math. Soc.* **26** (1979) A-451.

E. Grosswald, Arithmetic progressions of arbitrary length and consisting only of primes and almost primes, *J. reine angew. Math.* **317** (1980) 200–208.

Emil Grosswald and Peter Hagis, Arithmetic progressions consisting only of primes, *Math. Comput.* **33** (1979) 1343–1352; *MR* **80k**: 10054.

H. Halberstam, D. R. Heath-Brown, and H.-E. Richert, On almost-primes in short intervals, *Proc. Durham Sympos. Anal. Number Theory* (to appear).

D. R. Heath-Brown, Almost-primes in arithmetic progressions and short intervals, *Math. Proc. Cambridge Philos. Soc.*, **83** (1978) 357–375; *MR* **58** #10789.
Edgar Karst, 12–16 primes in arithmetical progression, *J. Recreational Math.* **2** (1969) 214–215.
E. Karst, Lists of ten or more primes in arithmetical progressions, *Scripta Math.* **28** (1970) 313–317.
E. Karst and S. C. Root, Teilfolgen von Primzahlen in arithmetischer Progression, *Anz. Oesterreich. Akad. Wiss. Math.-Naturwiss. Kl.* 1972, 19–20 (see also 178–179).
Carl Pomerance, The prime number graph, *Math. Comput.* **33** (1979) 399–408; MR **80d**:10013.
W. Sierpiński, Remarque sur les progressions arithmétiques, *Colloq. Math.* **3** (1955) 44–49.
Sol Weintraub, Seventeen primes in arithmetic progression, *Math. Comput.* **31** (1977) 1030.
S. Weintraub, Primes in arithmetic progression, *BIT* **17** (1977) 239–243.
K. Zarankiewicz, Problem 117, *Colloq. Math.* **3** (1955) 46, 73.

**A6.** It has even been conjectured that there are arbitrarily long arithmetic progressions of *consecutive* primes, such as

$$251, 257, 263, 269 \quad \text{and} \quad 1741, 1747, 1753, 1759.$$

Jones et al discovered the sequence $10^{10} + 24493 + 30k$ $(0 \le k \le 4)$ of five consecutive primes, and Lander and Parkin, soon after, found six such primes, $121174811 + 30k$ $(0 \le k \le 5)$. They also established that $9843019 + 30k$ $(0 \le k \le 4)$ is the least progression of five terms, that there are 25 others less than $3 \times 10^8$, but no others of length six.

It is not known if there are infinitely many sets of three *consecutive* primes in arithmetic progression, but S. Chowla has demonstrated this without the restriction to consecutive primes.

S. Chowla, There exists an infinity of 3-combinations of primes in A.P., *Proc. Lahore Philos. Soc.* **6** no. 2 (1944) 15–16; *MR* **7**, 243.
P. Erdös, and A. Rényi, Some problems and results on consecutive primes, *Simon Stevin*, **27** (1950) 115–125; *MR* **11**, 644.
M. F. Jones, M. Lal, and W. J. Blundon, Statistics on certain large primes, *Math. Comput.* **21** (1967) 103–107; *MR* **36** #3707.
L. J. Lander and T. R. Parkin, Consecutive primes in arithmetic progression, *Math. Comput.* **21** (1967) 489.

**A7.** A common method for proving that $p$ is a prime involves the factorization of $p - 1$. If $p - 1 = 2q$, where $q$ is another prime, the size of the problem has only been reduced by a factor of 2, so it's interesting to observe **Cunningham chains** of primes with each member one more than twice the previous one. D. H. Lehmer found that there were just three such chains of 7 primes with least member $< 10^7$:

1122659, 2245319, 4490639, 8981279, 17962559, 35925119, 71850239
2164229, 4328459, 8656919, 17313839, 34627679, 69255359, 138510719
2329469, 4658939, 9317879, 18635759, 37271519, 74543039, 149086079

and two others with least members 10257809 and 10309889. The factorization of $p + 1$ can also be used to prove that $p$ is prime. Lehmer also found seven chains of length 7 based on $p + 1 = 2q$. The first three have least members 16651, 67651, and 165901. No chain of length 8 of either kind is known. Lehmer estimates that they may be found with least members around $10^9$, about once in every 6 or 7 million trials.

D. H. Lehmer, Tests for primality by the converse of Fermat's theorem, *Bull. Amer. Math. Soc.* **33** (1927) 327–340.
D. H. Lehmer, On certain chains of primes, *Proc. London Math. Soc.* **14A** (Littlewood 80 volume, 1965) 183–186.

**A8.** There are many problems concerning the gaps between consecutive primes. Write $d_n = p_{n+1} - p_n$ so that $d_1 = 1$ and all other $d_n$ are even. How large and how small can $d_n$ be? Rankin has shown that

$$d_n > \frac{c \ln n \ln \ln n \ln \ln \ln \ln n}{(\ln \ln \ln n)^2}$$

for infinitely many $n$ and Erdös offers \$10,000 for a proof or disproof that the constant $c$ can be taken arbitrarily large. Rankin's best value is $c = e^\gamma$ where $\gamma$ is Euler's constant.

A very famous conjecture is the Twin Prime Conjecture, that $d_n = 2$ infinitely often. Conjecture B of Hardy and Littlewood (cf. A1) is that $P_k(n)$, the number of pairs of primes less than $n$ and differing by an even number $k$, is given asymptotically by

$$P_k(n) \sim \frac{2cn}{(\ln n)^2} \prod \left(\frac{p-1}{p-2}\right)$$

where the product is taken over all odd prime divisors of $k$ (and so is empty and taken to be 1 when $k$ is a power of 2) and $c = \prod(1 - 1/(p-1)^2)$ taken over all odd primes, so that $2c \approx 1.32032$. The large twin primes $9 \times 2^{211} \pm 1$ were discovered by the Lehmers and independently by Riesel. Recently Crandall and Penk have found twin primes with 64, 136, 154, 203 and 303 digits, Williams found $156 \times 5^{202} \pm 1$ and Baillie $297 \times 2^{546} \pm 1$. Even more recently, Atkin and Rickert have found pairs of twin primes

$$694513810 \times 2^{2304} \pm 1 \quad \text{and} \quad 1159142985 \times 2^{2304} \pm 1$$

They observe that the second pair is larger than the first.

Bombieri and Davenport have shown that

$$\liminf \frac{d_n}{\ln n} < \frac{2 + \sqrt{3}}{8} \approx 0.46650$$

(no doubt the real answer is zero; of course, the truth of the Twin Prime Conjecture would imply this). Huxley has shown that

$$d_n < p_n^{7/12 + \varepsilon}$$

and Heath-Brown and Iwaniec have recently improved the exponent to 11/20. Cramér proved, using the Riemann Hypothesis, that

$$\sum_{n<x} d_n^2 < cx(\ln x)^4$$

Erdös conjectures that the right-hand side should be $cx(\ln x)^2$, but thinks that there is no hope of a proof. The Riemann Hypothesis implies that $d_n < p_n^{1/2+\varepsilon}$.

Shanks has given a heuristic argument which supports the conjecture that if $p(g)$ is the first prime that follows a gap of $g$ or more composite numbers, then $\ln p(g) \sim \sqrt{g}$. Lehmer tabulated information for all primes $<37 \times 10^6$ and observed the gap $g = 209$ between the primes 20831323 and 20831533. Lander and Parkin extended the work for all $g < 314$ and Brent continued to $g < 534$. The entries for $g = 381$ and 651 in column $p_{n+1}$ of Table 2 are values of $p(g)$. Weintraub has found a gap of 653 near $1.1 \times 10^{16}$.

Table 2. Some Large Gaps between Consecutive Primes.

| $g$ | $p_n$ | $p_{n+1}$ | discoverers |
|---|---|---|---|
| 209 | 20831323 | 20831533 | Lehmer |
| 219 | 47326693 | 47326913 | Parkin |
| 221 | 122164747 | 122164969 | Lander & Parkin |
| 233 | 189695659 | 189695893 | Lander & Parkin |
| 281 | 436273009 | 436273291 | Lander & Parkin |
| 291 | 1453168141 | 145318433 | Lander & Parkin |
| 381 | 10726904659 | 10726905041 | Lander & Parkin |
| 463 | 42652618343 | 42652618807 | Brent |
| 533 | 614487453523 | 614487454057 | Brent |
| 601 | 1968188556461 | 1968188557063 | Brent |
| 651 | 2614941710599 | 2614941711251 | Brent |

Chen Jing-Run has shown that, for $x$ large enough, there is always a number with at most two prime factors in the interval $[x - x^\alpha, x]$ for any value of $\alpha \geq 0.477$.

A. O. L. Atkin and N. W. Rickert, On a larger pair of twin primes. Abstract 79T-A132, *Notices Amer. Math. Soc.* **26** (1979) A-373.
Robert Baillie, New primes of the form $k \cdot 2^n + 1$, *Math. Comput.* **33** (1979) 1333–1336.
E. Bombieri and H. Davenport, Small differences between prime numbers, *Proc. Roy. Soc.* Ser. A, **293** (1966) 1–18; *MR* **33** #7314.
Richard P. Brent, The first occurrence of large gaps between successive primes, *Math. Comput.* **27** (1973) 959–963; *MR* **48** #8360 (and see **35** (1980) 1435–1436).
J. H. Cadwell, Large intervals between consecutive primes, *Math. Comput.* **25** (1971) 909–913.
Chen Jing-Run, On the distribution of almost primes in an interval II, *Sci. Sinica* **22** (1979) 253–275; *Zbl* 408.10030.
R. J. Cook, On the occurrence of large gaps between prime numbers, *Glasgow Math. J.* **20** (1979) 43–48; *MR* **80e**:10034.
Harald Cramér, On the order of magnitude of the difference between consecutive prime numbers, *Acta Arith.* **2** (1937) 23–46.

R. E. Crandall and M. A. Penk, A search for large twin prime pairs, *Math. Comput.* **33** (1979) 383–388; *MR* **81a**:10010.
D. R. Heath-Brown, The difference between consecutive primes, *J. London Math. Soc.* (2) **18** (1978) 7–13, **19** (1979) 207–220; **20** (1979) 177–178; *MR* **58** #10787; **80k**: 10041; **81f**: 10055.
D. R. Heath-Brown and H. Iwaniec, On the difference between consecutive primes, *Inventiones Math.* **55** (1979) 49–69; *MR* **81h**: 10064; see also *Bull. Amer. Math. Soc.* (N.S.) **1** (1979) 758–760; *MR* **80d**: 10064.
Martin N. Huxley, The difference between consecutive primes, *Proc. Symp. Pure Math. Amer. Math. Soc.* **24** (Analytic Number Theory, St. Louis, 1972) 141–145; *MR* **50** #9816.
Martin N. Huxley, On the difference between consecutive primes, *Invent. Math.* **15** (1972) 164–170.
M. N. Huxley, A note on large gaps between prime numbers, *Acta Arith.* **38** (1980) 63–68.
M. N. Huxley, Small differences between consecutive primes I, *Mathematika*, **20** (1973) 229–232; *MR* **50** #4509; II, **24** (1977) 142–152; *MR* **57** #5925
Aleksandar Ivić, On sums of large differences between consecutive primes, *Math. Ann.* **241** (1979) 1–9; *MR* **80i**:10057.
Henryk Iwaniec and Matti Jutila, Primes in short intervals, *Ark. Mat.* **17** (1979) 167–176; *Zbl.* 408.10029.
L. J. Lander and T. R. Parkin, On first appearance of prime differences, *Math. Comput.* **21** (1967) 483–488; *MR* **37** #6237.
D. H. Lehmer, Table concerning the distribution of primes up to 37 millions, 1957, deposited in UMT file, reviewed in *Math. Tables Aids Comput.* **13** (1959) 56–57.
R. A. Rankin, The difference between consecutive primes, *J. London Math. Soc.* **13** (1938) 242–247; II *Proc. Cambridge Philos. Soc.* **36** (1940) 255–266; *MR* **1**-292; III *J. London Math. Soc.* **22** (1947) 226–230; *MR* **9**-498; IV *Proc. Amer. Math. Soc.* **1** (1950) 143–150; *MR* **11**-644; V *Proc. Edinburgh Math. Soc.* (2) **13** (1962–63) 331–332; *MR* **28** #3978.
H. Riesel, Lucasian criteria for the primality of $N = h \cdot 2^n - 1$, *Math. Comput.* **23** (1969) 869–875.
Daniel Shanks, On maximal gaps between successive primes, *Math. Comput.* **18** (1964) 646–651; *MR* **29** #4745.
Sol Weintraub, A large prime gap, *Math. Comput.* **36** (1981) 279.
H. C. Williams, Primality testing on a computer, *Ars Combinatoria*, **5** (1978) 172–185.
H. C. Williams and C. R. Zarnke, A report on prime numbers of the form $M = (6a + 1)2^{2m-1} - 1$ and $M' = (6a - 1)2^{2m} - 1$, *Math. Comput.* 22 (1968) 420–422.
D. Wolke, Grosse Differenzen zwischen aufeinanderfolgenden Primzahlen, *Math. Ann.* **218** (1975) 269–271; *MR* **56** #11930.

**A9.** A conjecture more general than the Twin Prime Conjecture is that there are infinitely many sets of primes of any given pattern, provided that there are no congruence relations which rule them out. It seems likely, for example, that there are infinitely many triples of primes $\{6k-1, 6k+1, 6k+5\}$ and $\{6k + 1, 6k + 5, 6k + 7\}$. This would be even harder to settle than the Twin Prime Conjecture, but its plausibility is of interest, since Hensley and Richards have shown that it is incompatible with the well-known conjecture (also due to Hardy and Littlewood)

$$¿¿¿ \qquad \pi(x + y) \leq \pi(x) + \pi(y) \qquad ???$$

for all integers $x, y \geq 2$. We've put more queries than usual by this, since it is very likely to be false. Indeed, there's some hope of finding values of $x$

and $y$ which contradict it. However there's an alternative conjecture,

$$¿ \quad \pi(x + y) \le \pi(x) + 2\pi(y/2) \quad ?$$

that the Hensley-Richards method doesn't comment on.

H. F. Smith noted that the pattern 11, 13, 17, 19, 23, 29, 31, 37 is repeated at least three times, starting with the primes 15760091, 25658841 and 93625991. In none of these cases is the number corresponding to 41 a prime, although $n - 11$, $n - 13 \ldots, n - 41$ are all primes for $n = 88830$ and 855750. In neither case is $n - 43$ prime, and John Leech observes that it is an unsolved problem to find 33 consecutive numbers greater than 11 which include 10 primes. More generally, find a set of $n$ numbers which congruence conditions do not preclude from containing more than $\pi(n)$ primes.

Paul Erdös and Ian Richards, Density functions for prime and relatively prime numbers, *Monatsh. Math.* **83** (1977) 99–112; Zbl. 355.10034.

Douglas Hensley and Ian Richards, On the incompatibility of two conjectures concerning primes, *Proc. Symp. Pure Math. Amer. Math. Soc.* **24** (Analytic Number Theory, St. Louis, 1972) 123–127.

Ian Richards, On the incompatibility of two conjectures concerning primes; a discussion of the use of computers in attacking a theoretical problem, *Bull. Amer. Math. Soc.* **80** (1974) 419–438.

Herschel F. Smith, On a generalization of the prime pair problem, *Math. Tables Aids Comput.* **11** (1957) 249–254.

## A10.

Define $d_n^k$ by $d_n^1 = d_n$ and $d_n^{k+1} = |d_{n+1}^k - d_n^k|$, that is, the successive absolute differences of the sequence of primes (Figure 2). N. L. Gilbreath conjectured that $d_1^k = 1$ for all $k$. This has been verified for $k < 63419$, that is, for all primes $<792722$, by Killgrove and Ralston.

1 2 3 4 5 6 7 8 9 10 11 12 13 14 15 16 17 18 19 20 21 22 23 24
2 3 5 7 11 13 17 19 23 29 31 37 41 43 47 53 59 61 67 71 73 79 83 89
1 2 2 4 2 4 2 4 6 2 6 4 2 4 6 6 2 6 4 2 6 4 6
1 0 2 2 2 2 2 2 4 4 2 2 2 2 0 4 4 2 2 4 2 2
1 2 0 0 0 0 0 2 0 2 0 0 0 2 4 0 2 0 2 2 0
1 2 0 0 0 0 2 2 2 2 0 0 2 2 4 2 2 2 0 2
1 2 0 0 0 2 0 0 0 2 0 2 0 2 2 0 0 2 2
1 2 0 0 2 2 0 0 2 2 2 2 2 0 2 0 2 0
1 2 0 2 0 2 0 2 0 0 0 0 2 2 2 2 2
1 2 2 2 2 2 2 2 0 0 0 2 0 0 0 0
1 0 0 0 0 0 0 2 0 0 2 2 0 0 0

Figure 2. Successive Absolute Differences of the Sequence of Primes.

It has been suggested that writers on this topic have been unduly mystical about it, and that it has nothing to do with primes as such, but will be true for any sequence consisting of 2 and odd numbers, which increases at a "reasonable" rate and has gaps of "reasonable" size.

R. B. Killgrove and K. E. Ralston, On a conjecture concerning the primes, *Math Tables Aids Comput.* **13** (1959) 121–122; *MR* **21** #4943.

**A11.** Since the proportion of primes gradually decreases, albeit somewhat erratically, $d_m < d_{m+1}$ infinitely often and Erdös and Turán have shown that the same is true for $d_n > d_{n+1}$. They have also shown that the values of $n$ for which $d_n > d_{n+1}$ have positive lower density, but it is not known if there are infinitely many decreasing or increasing sets of *three* consecutive values of $d_n$. If there were not, then there is an $n_0$ so that for every $i$ and $n > n_0$, we have $d_{n+2i} > d_{n+2i+1}$ and $d_{n+2i+1} < d_{n+2i+2}$. Erdös offers \$100 for a proof that such an $n_0$ does not exist.

P. Erdös, On the difference of consecutive primes, *Bull. Amer. Math. Soc.* **54** (1948) 885–889; *MR* **10**, 235.

P. Erdös and P. Turán, On some new questions on the distribution of prime numbers, *Bull. Amer. Math. Soc.* **54** (1948) 371–378; *MR* **9**, 498.

**A12.** Pomerance, Selfridge, and Wagstaff call an odd composite $n$ for which $a^{n-1} \equiv 1 \pmod n$ a **pseudoprime to base** $a$ (psp($a$)). This usage is introduced to avoid the clumsy "composite pseudoprime" which appears throughout the literature. Odd composite $n$ which are psp($a$) for every $a$ prime to $n$ are **Carmichael numbers**. An odd composite $n$ is an **Euler pseudoprime to base** $a$ (epsp($a$)) if $(a, n) = 1$ and $a^{(n-1)/2} \equiv (\frac{a}{n}) \pmod n$, where $(\frac{a}{n})$ is the Jacobi symbol (see F5). Finally, an odd composite $n$ with $n - 1 = d \times 2^s$, $d$ odd, is a **strong pseudoprime to base** $a$ (spsp($a$)) if $a^d \equiv 1 \pmod n$ (otherwise $a^{d \times 2^r} \equiv -1 \pmod n$ for some $r$, $0 \le r < s$). These definitions are illustrated by a Venn diagram (Figure 3) which displays the smallest member of each set.

The following values of $P_2(x)$, $E_2(x)$, $S_2(x)$, and $C(x)$—the numbers of psp(2), epsp(2), spsp(2), and Carmichael numbers less than $x$, respectively—were given by Pomerance, Selfridge, and Wagstaff:

| $x$ | $10^3$ | $10^4$ | $10^5$ | $10^6$ | $10^7$ | $10^8$ | $10^9$ | $10^{10}$ | $2 \cdot 5 \times 10^{10}$ |
|---|---|---|---|---|---|---|---|---|---|
| $P_2(x)$ | 3 | 22 | 78 | 245 | 750 | 2057 | 5597 | 14884 | 21853 |
| $E_2(x)$ | 1 | 12 | 36 | 114 | 375 | 1071 | 2939 | 7706 | 11347 |
| $S_2(x)$ | 0 | 5 | 16 | 46 | 162 | 488 | 1282 | 3291 | 4842 |
| $C(x)$ | 1 | 7 | 16 | 43 | 105 | 255 | 646 | 1547 | 2163 |

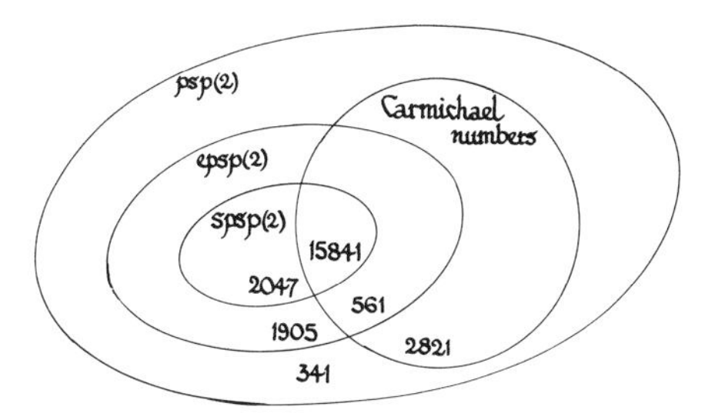


Figure 3. Relationship of Sets of psps with Least Element in Each Set.

Lehmer and Erdös showed that

$$c_1 \ln x < P_2(x) < x \exp\{-c_2(\ln x \ln\ln x)^{1/2}\}$$

and Pomerance has recently improved these bounds to

$$\exp\{(\ln x)^{5/14}\} < P_2(x) < x \exp\{(-\ln x \ln\ln\ln x)/2 \ln\ln x\}$$

and has a heuristic argument that the true estimate is the upper bound with the 2 omitted.

There are also even psp(2). Lehmer found $161038 = 2 \times 73 \times 1103$ and Beeger showed there are infinitely many. If $F_n$ is the Fermat number $2^{2^n} + 1$, Cipolla showed that $F_{n_1}F_{n_2} \cdots F_{n_k}$ is psp(2) if $k > 1$ and $n_1 < n_2 < \cdots < n_k < 2^{n_1}$.

If $P_n^{(a)}$ is the $n$th psp($a$), Szymiczek has shown that $\sum 1/P_n^{(2)}$ is convergent, while Mąkowski has shown that $\sum 1/\ln P_n^{(a)}$ is divergent. Rotkiewicz has a booklet on pseudoprimes which contains 58 problems and 20 conjectures.


N. G. W. H. Beeger, On even numbers $m$ dividing $2^m - 2$, *Amer. Math. Monthly*, **58** (1951) 553–555; *MR* **13**, 320.

R. D. Carmichael, On composite numbers $P$ which satisfy the Fermat congruence $a^{P-1} \equiv 1 \bmod P$, *Amer. Math. Monthly* **19** (1912) 22–27.

M. Cipolla, Sui numeri composti $P$ che verificiano la congruenza di Fermat, $a^{P-1} \equiv 1 \pmod{P}$, *Annali di Matematica* **9** (1904) 139–160.

P. Erdös, On the converse of Fermat's theorem, *Amer. Math. Monthly* **56** (1949) 623–624; *MR* **11**, 331.

P. Erdös, On almost primes, *Amer. Math. Monthly* **57** (1950) 404–407; *MR* **12**, 80.

D. H. Lehmer, On the converse of Fermat's theorem, *Amer. Math. Monthly* **43** (1936) 347–354; II, **56** (1949) 300–309; *MR* **10**, 681.

Andrzej Mąkowski, On a problem of Rotkiewicz on pseudoprime numbers, *Elem. Math.* **29** (1974) 13.

A. Mąkowski and A. Rotkiewicz, On pseudoprime numbers of special form, *Colloq. Math.* 20 (1969) 269–271; *MR* **39** #5458.

Carl Pomerance, *Illinois J. Math.* (to appear).

Carl Pomerance, On the distribution of pseudoprimes, *Math. Comput.* (to appear).

Carl Pomerance, John L. Selfridge, and Samuel S. Wagstaff, The pseudoprimes to $25 \cdot 10^9$, *Math. Comput.* **35** (1980) 1003–1026.

A. Rotkiewicz, *Pseudoprime Numbers and their Generalizations*, Student Association of the Faculty of Sciences, Univ. of Novi Sad, 1972; *MR* **48** #8373; *Zbl.* 324.10007.

A. Rotkiewicz, Sur les diviseurs composés des nombres $a^n - b^n$, *Bull. Soc. Roy. Sci. Liège* **32** (1963) 191–195; *MR* **26** #3645.

A. Rotkiewicz, Sur les nombres pseudopremiers de la forme $ax + b$, *Comptes Rendus Acad. Sci. Paris* **257** (1963) 2601–2604; *MR* **29** #61.

A. Rotkiewicz, Sur les formules donnant des nombres pseudopremiers, *Colloq. Math.* **12** (1964) 69–72; *MR* **29** #3416.

A. Rotkiewicz, Sur les nombres pseudopremiers de la forme $nk + 1$, *Elem. Math.* **21** (1966) 32–33; *MR* **33** #112.

K. Szymiczek, On prime numbers $p$, $q$ and $r$ such that $pq$, $pr$ and $qr$ are pseudoprimes, *Colloq. Math.* **13** (1964–65) 259–263; *MR* **31** #4757.

K. Szymiczek, On pseudoprime numbers which are products of distinct primes, *Amer. Math. Monthly* **74** (1967) 35–37; *MR* **34** #5746.


**A13.** The Carmichael numbers (psp($a$) for all $a$ prime to $n$) must have at least three prime factors, e.g. $561 = 3 \times 11 \times 17$. It is not known if there is an infinite number of Carmichael numbers, but Erdös conjectures that $(\ln C(x))/\ln x$ tends to 1 as $x$ tends to infinity. He improves a result of Knödel to show that

$$C(x) < x \exp\{-c \ln x \ln\ln\ln x/\ln\ln x\}$$

while Pomerance, Selfridge and Wagstaff (see A12) prove this with $c = 1 - \varepsilon$ and give a heuristic argument supporting the conjecture that the reverse inequality holds with $c = 2 + \varepsilon$.

J. R. Hill has found the large Carmichael number $pqr$ where $p = 5 \times 10^{19} + 371$, $q = 10^{20} + 741$ and $r = 1 + (p - 1)(q + 2)/433$.


Robert Baillie and Samuel S. Wagstaff, Lucas pseudoprimes, *Math. Comput.* **35** (1980) 1391–1417.

P. Erdös, On pseudoprimes and Carmichael numbers, *Publ. Math. Debrecen* **4** (1956) 201–206; MR **18**, 18.

Jay Roderick Hill, Large Carmichael numbers with three prime factors, Abstract 79T-A136, *Notices Amer. Math. Soc.* **26** (1979) A-374.

W. Knödel, Eine obere Schranke für die Anzahl der Carmichaelschen Zahlen kleiner als $x$, *Arch. Math.* **4** (1953) 282–284; *MR* **15** 289.

D. H. Lehmer, Strong Carmichael numbers, *J. Austral. Math. Soc.* Ser. A **21** (1976) 508–510.

A. J. van der Poorten and A. Rotkiewicz, On strong pseudoprimes in arithmetic progressions, *J. Austral. Math. Soc.* Ser. A, **29** (1980) 316–321.

S. S. Wagstaff, Large Carmichael numbers, *Math. J. Okayama Univ.* **22** (1980) 33–41.

H. C. Williams, On numbers analogous to Carmichael numbers, *Canad. Math. Bull.* **20** (1977) 133–143.

M. Yorinaga, Numerical computation of Carmichael numbers, *Math. J. Okayama Univ.* **20** (1978) 151–163. *MR* **80d**:10026.


## A14.

Erdös and Straus call the prime $p_n$ **good** if $p_n^2 > p_{n-i}p_{n+i}$ for all $i$ $(1 \le i \le n - 1)$; for example, 5, 11, 17, and 29. Pomerance used the "prime number graph" (see A5) to show that there are infinitely many good primes. He asks the following questions. Is it true that the set of $n$ for which $p_n$ is good has density 0? Are there infinitely many $n$ with $p_n p_{n+1} > p_{n-i}p_{n+1+i}$ for all $i$ $(0<i<n)$? Are there infinitely many $n$ with $p_n + p_{n+1} < p_{n-i} + p_{n+1+i}$ for all $i$ $(0 < i < n)$? Does the set of $n$ for which $2p_n < p_{n-i} + p_{n+i}$ for all $i$ $(0 < i < n)$ have density 0? (Pomerance proved there are infinitely many such $n$.) Is $\limsup(\min_{0<i<n}(p_{n-i} + p_{n+i}) - 2p_n) = \infty$?

## A15.

Erdös, in a letter dated 79:10:31 observes that $3 \times 4 \equiv 5 \times 6 \times 7 \equiv 1 \pmod{11}$ and asks for the least prime $p$ such that there are integers $a$, $k_1$, $k_2$, $k_3$ and

$$\prod_{i=1}^{k_1}(a+i) \equiv \prod_{i=1}^{k_2}(a+k_1+i) \equiv \prod_{i=1}^{k_3}(a+k_1+k_2+i) \equiv 1 \pmod{p}.$$

He suggests that such primes $p$ exist for any number of such congruent products.

**A16.** Prime numbers can be defined in fields other than the rational field. In the complex number field they are called **Gaussian primes**. Many problems on ordinary primes can be reformulated for Gaussian primes.

**Gaussian integers**, $a + bi$ where $a$, $b$ are integers and $i^2 = -1$, behave like ordinary integers in the sense that there is **unique factorization** (apart from order, **units** $(\pm 1, \pm i)$ and **associates**; the associates of 7, for example, are 7, $-7$, $7i$, and $-7i$). Primes of the form $4k - 1$ are still primes $(3, 7, 11, 19, 23, \ldots)$ but the other ordinary primes can be factored into Gaussian primes:

$$2 = (1 + i)(1 - i), \qquad 5 = (2 + i)(2 - i) = -(2i - 1)(2i + 1), \text{ etc.}$$

$$13 = (2 + 3i)(2 - 3i), \qquad 17 = (4 + i)(4 - i), \qquad 29 = (5 + 2i)(5 - 2i), \ldots .$$

The Gaussian primes $\pm 1 \pm i$, $\pm 2 \pm i$, $\pm 3$, $\pm 3i$, $\pm 2 \pm 3i$, $\pm 4 \pm i$, $\pm 5 \pm 2i, \ldots$ make a pleasing pattern (Figure 4) when drawn on an Argand diagram, which has been used for tiling floors and weaving tablecloths.

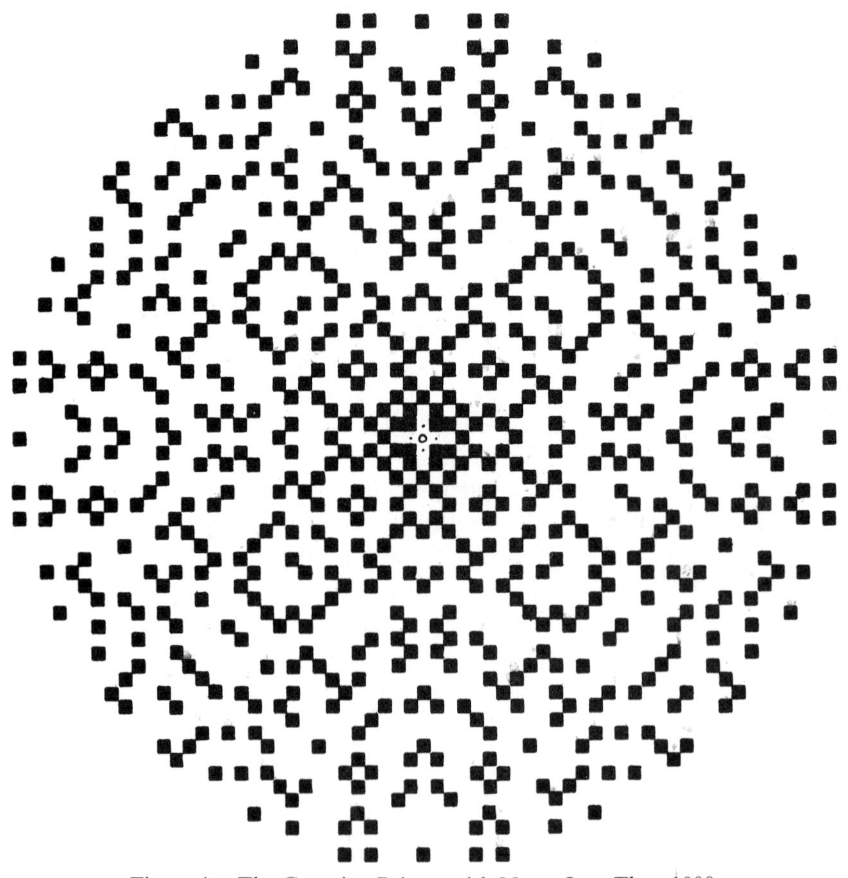

Figure 4. The Gaussian Primes with Norm Less Than 1000.

Motzkin and Gordon asked if one can "walk" from the origin to infinity using the Gaussian primes as "stepping stones," and taking steps of bounded length. Presumably not. Jordan and Rabung have shown that steps of length at least 4 are necessary.

The **Eisenstein integers** $a + b\omega$, where $a$, $b$ are integers and $\omega$ is a complex cube root of unity, $\omega^2 + \omega + 1 = 0$, also enjoy unique factorization. The primes again form a pattern, this time with hexagonal symmetry, because there are six units, $\pm 1$, $\pm\omega$, $\pm\omega^2$. The prime 2 and those of form $6k - 1$ $(5, 11, 17, 23, 29, 41, \ldots)$ are still Eisenstein primes, but 3 and those of form $6k + 1$ can be factored:

$$3=(1-\omega)(1-\omega^2), \qquad 7=(2-\omega)(2-\omega^2), \qquad 13=(3-\omega)(3-\omega^2),$$
$$19=(3-2\omega)(3-2\omega^2), \qquad 31=(5-\omega)(5-\omega^2), \qquad 37=(4-3\omega)(4-3\omega^2), \ldots .$$

The **Eisenstein primes**, in the field containing the complex cube roots of unity, are depicted in Figure 5. Corresponding problems can be formulated for these.

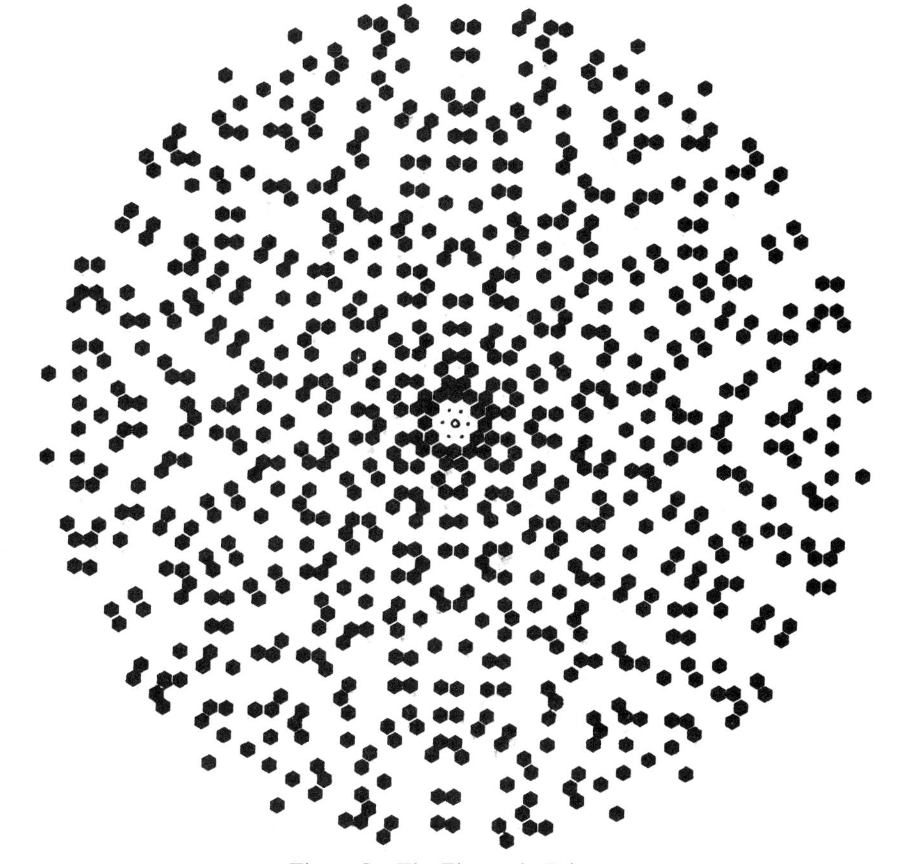

Figure 5. The Eisenstein Primes.

J. H. Jordan and J. R. Rabung, A conjecture of Paul Erdös concerning Gaussian primes, *Math. Comput.* **24** (1970) 221–223.

## A17.

Perhaps the philosopher's stone of number theory is a formula for $p_n$, or for $\pi(x)$, or for a necessary and sufficient condition for primality. Wilson's theorem seems to be unique, but even that is useless for computation. C. P. Willans and C. P. Wormell used it to give formulas which use only elementary functions, but which are too clumsy to print here. Matjasevič and other logicians have used Wilson's theorem to solve Hilbert's tenth problem. Is there any alternative to Wilson's theorem?

Sierpiński observes that it follows from Fermat's theorem that if $p$ is prime, then $p$ divides

$$1^{p-1} + 2^{p-1} + \cdots + (p-1)^{p-1} + 1.$$

Is the converse true? Giuga has verified it for $p \le 10^{1000}$.

See also B31 for Wolstenholme's theorem.

L. E. Dickson, *History of the Theory of Numbers*, G. E. Stechert & Co, New York, 1934, Vol. I, Chap. XVIII.
G. Giuga, Sopra alcune proprietà caratteristiche dei numeri primi, *Period. Math.* (4) **23** (1943) 12–27; *MR* **8**, 11.
Giuseppe Giuga, Su una presumibile proprietà caratteristica dei numeri primi, *Ist. Lombardo Sci. Lett. Rend. Cl. Sci. Mat. Nat.* (3) **14** **(83)** (1950) 511–528; *MR* **13**, 725.
W. Sierpiński, *Elementary Number Theory*, p. 205.
C. P. Willans, On formulae for the $N$th prime number, *Math. Gaz.* **48** (1964) 413–415.
C. P. Wormell, Formulae for primes, *Math. Gaz.* **51** (1967) 36–38.

## A18.

Erdös and Selfridge classify the primes as follows: $p$ is in class 1 if the only prime divisors of $p + 1$ are 2 or 3, and $p$ is in class $r$ if every prime factor of $p + 1$ is in class $\le r - 1$, with equality for at least one prime factor. For example:

class 1: 2 3 5 7 11 17 23 31 47 53 71 107 127 191 431 647 863 971 . . .

class 2: 13 19 29 41 43 59 61 67 79 83 89 97 101 109 131 137 139 149 167 179 197 199 211 223 229 239 241 251 263 269 271 281 283 293 307 317 319 359 367 373 377 383 419 439 449 461 467 499 503 509 557 563 577 587 593 599 619 641 643 659 709 719 743 751 761 769 809 827 839 881 919 929 953 967 979 991 . . .

class 3: 37 103 113 151 157 163 173 181 193 227 233 257 277 311 331 337 347 353 379 389 397 401 409 421 457 463 467 487 491 521 523 541 547 571 601 607 613 631 653 683 701 727 733 773 787 811 821 829 853 857 859 877 883 911 937 947 983 997 . . .

class 4: 73 313 443 617 661 673 677 691 739 757 823 887 907 941 977 ...

class 5: 1021 1321 1381 ...

It's easy to prove that the number of primes in class $r$, not exceeding $n$, is $o(n^\varepsilon)$ for every $\varepsilon > 0$ and all $r$. Prove that there are infinitely many primes in each class. If $p_1^{(r)}$ denotes the least prime in class $r$, so that $p_1^{(1)} = 2$, $p_1^{(2)} = 13$, $p_1^{(3)} = 37$, $p_1^{(4)} = 73$, and $p_1^{(5)} = 1021$, then Erdös thought that $(p_1^{(r)})^{1/r} \to \infty$, while Selfridge thought it quite likely to be bounded.

A similar classification arises if $p - 1$ replaces $p + 1$:

class 1: 2 3 5 7 13 17 19 37 73 97 109 ...

class 2: 11 29 31 41 43 53 61 71 79 101 103 113 127 131 137 149 151 157 ...

class 3: 23 59 67 83 89 107 ...

class 4: 47 139 ...

class 5: 283 ...

for which similar answers are to be expected.

P. Erdös, Problems in Number Theory and Combinatorics, *Congressus Numerantium XVIII*, Proc. 6th Conf. Numerical Math., Manitoba, 1976, 35–58 (esp. p. 53); *MR* **80e**:10005.

**A19.** Erdös conjectures that 7, 15, 21, 45, 75, and 105 are the only values of $n$ for which $n - 2^k$ is prime for all $k$ such that $2 \le 2^k < n$. Mientka and Weitzenkamp have verified this for $n < 2^{44}$. Vaughan has proved that there are not too many such numbers, less than $x \exp\{-(\ln x)^c\}$ of them less than $x$, but he was unable to show that there were less than $x^{1-\varepsilon}$.

Erdös also conjectures that for infinitely many $n$, all the integers $n - 2^k$, $1 \le 2^k < n$ are squarefree (cf. also F13).

Cohen and Selfridge ask for the least positive odd number *not* of the form $\pm p^a \pm 2^b$, where $p$ is a prime, $a \ge 0$, $b \ge 1$ and any choice of signs may be made. They observe that the number is greater than $2^{18}$, but at most

6120 6699060672 7677809211 5601756625 4819576161-
6319229817 3436854933 4512406741 7420946855 8999326569

Crocker proved that there are infinitely many odd integers *not* of the form $2^k + 2^l + p$, where $p$ is prime. Erdös asks if, for each $r$, there are infinitely many odd integers not the sum of a prime and $r$ or fewer powers of 2. Is their density positive? Do they contain an infinite arithmetic progression? In the opposite direction, Gallagher has proved that for every $\varepsilon > 0$ there is a sufficiently large $r$ so that the lower density of sums of primes with $r$ powers of 2 is greater than $1 - \varepsilon$.

Erdös also asks if there is an odd integer *not* of the form $2^k + s$ where $s$ is squarefree. There are connexions between some of these problems and covering congruences (F13).

Let $f(n)$ be the number of representations of $n$ as a sum $2^k + p$, and let $\{a_i\}$ be the sequence of values of $n$ for which $f(n) > 0$. Does the density of $\{a_i\}$ exist? Erdös showed that $f(n) > c \ln\ln n$ infinitely often, but could not decide if $f(n) = o(\ln n)$. He conjectures that $\limsup(a_{i+1} - a_i) = \infty$. This would follow if there are covering systems with arbitrarily large least moduli.

Carl Pomerance notes that for $n = 210$, $n - p$ is prime for all $p$, $7 < p < n$, and asks if there is any other such $n$.

Fred Cohen and J. L. Selfridge, Not every number is the sum or difference of two prime powers, *Math. Comput.* **29** (1975) 79–81.

R. Crocker, On the sum of a prime and of two powers of two, *Pacific J. Math.* **36** (1971) 103–107; *MR* **43** #3200.

P. Erdös, On integers of the form $2^r + p$ and some related problems, *Summa Brasil. Math.* **2** (1947–51) 113–123; *MR* **13**, 437.

Patrick X. Gallagher, Primes and powers of 2, *Inventiones Math.* **29** (1975) 125–142.

W. E. Mientka and R. C. Weitzenkamp, On $f$-plentiful numbers, *J. Combin. Theory* **7** (1969) 374–377.

A. de Polignac, Recherches nouvelles sur les nombres premiers, *C. R. Acad. Sci. Paris* **29** (1849) 397–401, 738–739.

R. C. Vaughan, Some applications of Montgomery's sieve, *J. Number Theory* **5** (1973) 64–79; *MR* **49** #7222.

# B. Divisibility

We will denote by $d(n)$ the number of positive divisors of $n$, by $\sigma(n)$ the sum of those divisors, and by $\sigma_k(n)$ the sum of their $k$th powers, so that $\sigma_0(n) = d(n)$ and $\sigma_1(n) = \sigma(n)$. We use $s(n)$ for the sum of the **aliquot parts** of $n$, i.e., the positive divisors of $n$ other than $n$ itself, so that $s(n) = \sigma(n) - n$.

Iterations of various arithmetic functions will be denoted, for example, by $s^k(n)$, which is defined by $s^0(n) = n$ and $s^{k+1}(n) = s(s^k(n))$ for $k \geq 0$.

We use the notation $d \mid n$ to mean that $d$ divides $n$, and $e \nmid n$ to mean that $e$ does not divide $n$. The notation $p^k \| n$ is used to imply that $p^k \mid n$ but $p^{k+1} \nmid n$. By $[m, n]$ we will mean the consecutive integers $m, m + 1, \ldots, n$.

**B1.** A **perfect number** is one such that $s(n) = n$. Euclid knew that $2^{p-1}(2^p - 1)$ was perfect if $2^p - 1$ is prime. For example, 6, 28, 496, etc.; see the list of Mersenne primes in A3. Euler showed that these were the only even perfect numbers.

The existence or otherwise of odd perfect numbers is one of the more notorious unsolved problems of number theory. Tuckerman, Hagis, Stubblefield, Buxton and Elmore have gradually pushed the bound below which an odd perfect number cannot exist, to above $10^{200}$, though there is some skepticism about the later proofs. Hagis and Chein have independently shown that an odd perfect number is divisible by at least 8 distinct primes and Muskat that it is divisible by a prime power $> 10^{12}$. Hagis and McDaniel show that the largest prime divisor is greater than 100110, and Pomerance that the next largest is greater than 138. Condict and Hagis have improved these bounds to 300000 and 1000 respectively. Pomerance has also shown that an odd perfect number with at most $k$ distinct factors is less than

$$(4k)^{(4k)^{2^{k^2}}}$$

M. Buxton and S. Elmore, An extension of lower bounds for odd perfect numbers, *Notices Amer. Math. Soc.* **23** (1976) A-55.

E. Z. Chein, PhD thesis, Pennsylvania State Univ. 1979.

E. Z. Chein, An odd perfect number has at least 8 prime factors, Abstract 79T-A102, *Notices Amer. Math. Soc.* **26** (1979) A-365.

G. L. Cohen, On odd perfect numbers, *Fibonacci Quart.* **16** (1978) 523–527; *MR* **80g**:10010; *Zbl.* 391.10008.

Graeme L. Cohen, On odd perfect numbers (II), Multiperfect numbers and quasiperfect numbers, *J. Austral. Math. Soc.* Ser. A **29** (1980) 369–384.

G. L. Cohen and M. D. Hendy, Polygonal supports for sequences of primes, *Math. Chronicle* **9** (1980) 120–136.

J. T. Condict, On an odd perfect number's largest prime divisor, senior thesis, Middlebury College, May, 1978.

G. G. Dandapat, J. L. Hunsucker, and Carl Pomerance, Some new results on odd perfect numbers, *Pacific J. Math.* **57** (1975) 359–364.

John A. Ewell, On the multiplicative structure of odd perfect numbers, *J. Number Theory* **12** (1980) 339–342.

O. Grün, Über ungerade vollkommene Zahlen, *Math. Zeit.* **55** (1952) 353–354.

Peter Hagis, A lower bound for the set of odd perfect numbers, *Math. Comput.* **27** (1973) 951–953.

P. Hagis, Every odd perfect number has at least 8 prime factors, *Notices Amer. Math. Soc.* **22** (1975) A-60.

Peter Hagis, Outline of a proof that every odd perfect number has at least eight prime factors, *Math. Comput.* **34** (1980) 1027–1032.

Peter Hagis, On the second largest prime factor of an odd perfect number. *Proc. Grosswald Conf., Lecture Notes in Mathematics*, Spring-Verlag, New York, (to be published).

Peter Hagis and Wayne McDaniel, On the largest prime divisor of an odd perfect number, *Math. Comput.* **27** (1973) 955–957; *MR* **48** #3855; II, *ibid.* **29** (1975) 922–924.

H.-J. Kanold, Untersuchungen über ungerade vollkommene Zahlen, *J. reine angew. Math.* **183** (1941) 98–109; *MR* **3**, 268.

P. J. McCarthy, Odd perfect numbers, *Scripta Math.* **23** (1957) 43–47.

Wayne McDaniel, On odd multiply perfect numbers, *Boll. Un. Mat. Ital.* (4) **3** (1970) 185–190; *MR* **41** #6764.

Wayne L. McDaniel and Peter Hagis, Some results concerning the non-existence of odd perfect numbers of the form $p^{\alpha}M^{2\beta}$, *Fibonacci Quart.* **13** (1975) 25–28.

Joseph B. Muskat, On divisors of odd perfect numbers, *Math. Comput.* **20** (1966) 141–144; *MR* **32** #4076.

Karl K. Norton, Remarks on the number of factors of an odd perfect number, *Acta Arith.* **6** (1961) 365–374 (and see **36** (1980) 163); *MR* **26** #4950.

M. Perisastri, A note on odd perfect numbers, *Math. Student* **26** (1958) 179–181.

C. Pomerance, Odd perfect numbers are divisible by at least seven distinct primes. *Acta Arith.* **25** (1974) 265–300; see also *Notices Amer. Math. Soc.* **19** (1972) A-622–623.

C. Pomerance, The second largest factor of an odd perfect number, *Math. Comput.* **29** (1975) 914–921.

Carl Pomerance, Multiply perfect numbers, Mersenne primes and effective computability, *Math. Ann.* **226** (1977) 195–206.

N. Robbins, The non-existence of odd perfect numbers with less than seven distinct prime factors, PhD dissertation, Polytech. Inst. Brooklyn, June 1972.

Hans Salié, Über abundante Zahlen, *Math. Nachr.* **9** (1953) 217–220.

C. Servais, Sur les nombres parfaits, *Mathesis* **8** (1888) 92–93.

Daniel Shanks, *Solved and Unsolved Problems in Number Theory*, 2nd ed. Chelsea, New York 1978, esp. p. 217.
Bryant Tuckerman, A search procedure and lower bound for odd perfect numbers, *Math. Comput.* **27** (1973) 943–949.

**B2.** Perhaps because they were frustrated by their failure to disprove the existence of odd perfect numbers, numerous authors have defined a number of closely related concepts and produced a raft of problems, many of which seem no more tractable than the original.

For a perfect number, $\sigma(n) = 2n$. If $\sigma(n) < 2n$, $n$ is called **deficient**; if $\sigma(n) > 2n$, **abundant**. If $\sigma(n) = 2n - 1$, $n$ has been called **almost perfect**. Powers of 2 are almost perfect; it is not known if any other numbers are. If $\sigma(n) = 2n + 1$, $n$ has been called **quasi-perfect**. Quasi-perfect numbers must be odd squares, but no one knows if there are any. Masao Kishore shows that $n > 10^{30}$ and that $\omega(n) \geq 6$, where $\omega(n)$ is the number of distinct prime factors of $n$. Hagis and Cohen have improved these results to $n > 10^{35}$ and $\omega(n) \geq 7$. Catteneo originally claimed to have proved that $3 \nmid n$, but Sierpiński and others have observed that his proof is fallacious. Kravitz, in a letter, makes a more general conjecture, that there are no numbers whose **abundance**, $\sigma(n) - 2n$, is an odd square. In this connexion Graeme Cohen writes that it is interesting that

$$\sigma(2^2 3^2 5^2) = 3(2^2 3^2 5^2) + 11^2,$$

and that if $\sigma(n) = 2n + k^2$ with $(n, k) = 1$ and $\omega(n) = 3$, then $105 | n$ or $165 | n$ and $n > 10^{500}$, $k > 10^{250}$. Later, relaxing the condition that $(n, k) = 1$, he finds the solutions

$$n = 2 \times 3^2 \times 238897^2, \qquad k = 3^2 \times 23 \times 1999$$

and

$$n = 2^2 \times 7^2 \times p^2, \quad \text{with } p = 53, 277, 541, 153941, 358276277,$$

and

$$k = 7 \times 29, 5 \times 7 \times 23, 5 \times 7 \times 43, 5 \times 7 \times 103 \times 113, 5 \times 7 \times 227 \times 229 \times 521$$

respectively. He verifies that the first of these last five is the smallest integer with odd square abundance. Erdös asks for a characterization of the large numbers for which $|\sigma(n) - 2n| < C$ for some constant $C$. For example, $n = 2^m$; are there other infinite families?

Sierpiński called a number **pseudoperfect** if it was the sum of *some* of its divisors. For example $20 = 1 + 4 + 5 + 10$. Erdös has shown that their density exists. Call a number **primitive abundant** if it is abundant, but all its proper divisors are deficient, and **primitive pseudoperfect** if it is pseudoperfect, but none of its proper divisors are. If the harmonic mean of all the divisors of $n$ is an integer, Pomerance called $n$ a **harmonic number**. A. and

E. Zachariou call these "Ore numbers" and they call primitive pseudoperfect numbers "irreducible semiperfect." They note that every multiple of a pseudoperfect number is pseudoperfect, and that the pseudoperfect numbers and the harmonic numbers both include the perfect numbers as a proper subset. The last result is due to Ore. All numbers $2^m p$ with $m \geq 1$ and $p$ a prime between $2^m$ and $2^{m+1}$ are primitive pseudoperfect, but there are such numbers not of this form, e.g., 770. There are infinitely many primitive pseudoperfect numbers that are not harmonic numbers. The smallest odd primitive pseudoperfect number is 945. Erdös can show that the number of odd primitive pseudoperfect numbers is infinite.

Garcia extended the list of harmonic numbers to include all 45 which are $<10^7$, and he found more than 200 larger ones. The least one, apart from 1 and the perfect numbers, is 140. Are any of them squares, apart from 1? Are there infinitely many of them? If so, find upper and lower bounds on the number of them that are $<x$. Kanold has shown that their desnity is zero, and Pomerance that a harmonic number of the form $p^a q^b$ ($p$ and $q$ primes) is an even perfect number. If $n = p^a q^b r^c$ is harmonic, is it even?

Which values does the harmonic mean take? Presumably not 4, 12, 16, 18, 20, 22, ...; does it take the value 23? Ore's own conjecture, that every Ore number is even, implies that there are no odd perfect numbers!

Bateman, Erdös, Pomerance, and Straus show that the set of $n$ for which $\sigma(n)/d(n)$ is an integer has density 1, that the set for which $\sigma(n)/d(n))^2$ is an integer has density $\frac{1}{2}$, and that the number of rationals $r \leq x$ of the form $\sigma(n)/d(n)$ is $o(x)$. They ask for an asymptotic formula for

$$\frac{1}{x}\sum 1$$

where the sum is taken over those $n \leq x$ for which $d(n)$ does *not* divide $\sigma(n)$. They also conjecture strongly that the integers $n$ for which $d(n)$ divides $s(n) = \sigma(n) - n$, have zero density, but they lack a straightforward proof.

Benkoski has called a number **weird** if it is abundant but not pseudoperfect. For example, 70 is not the sum of any subset of $1 + 2 + 5 + 7 + 10 + 14 + 35 = 74$. There are 24 primitive weird numbers less than a million: 70, 836, 4030, 5830, 7192, ... . Nonprimitive weird numbers include $70p$ with $p$ prime and $p > \sigma(70) = 144$; $836p$ with $p = 421, 487, 491$, or $p$ prime and $\geq 557$; $7192 \times 31$. Some large weird numbers were found by Kravitz, and Benkoski and Erdös showed that their density is positive. Here the open questions are: are there infinitely many primitive abundant numbers which are weird? Is every odd abundant number pseudoperfect (i.e., not weird)? Can $\sigma(n)/n$ be arbitrarily large for weird $n$? Benkoski and Erdös conjecture "no" in answer to the last question and Erdös offers \$10 and \$25 respectively for solutions to the last two questions.

Numbers have been called **multiply perfect** or ***k*-fold perfect** if $\sigma(n) = kn$. For example, ordinary perfect numbers are 2-fold perfect and 120 is 3-fold perfect. Dickson's *History* records a long interest in such numbers. The largest known value of $k$ is 8, for which Alan L. Brown gives three examples (and Franqui and Garcia two others) the smallest of which is $2 \times 3^{23} \times 5^9 \times 7^{12} \times 11^3 \times 13^3 \times 17^2 \times 19^2 \times 23 \times 29^2 \times 31^2 \times 37 \times 41 \times 53 \times 61 \times 67^2 \times 71^2 \times 73 \times 83 \times 89 \times 103 \times 127 \times 131 \times 149 \times 211 \times 307 \times 331 \times 463 \times 521 \times 683 \times 709 \times 1279 \times 2141 \times 2557 \times 5113 \times 6481 \times 10429 \times 20857 \times 110563 \times 599479 \times 16148168401$. No doubt $k$ can be as large as we wish, though Erdös conjectures that $k = o(\ln \ln n)$.

Minoli and Bear say that $n$ is $k$-**hyperperfect** if $n = 1 + k \sum d_i$, where the summation is taken over all proper divisors, $1 < d_i < n$, so $k\sigma(n) = (k+1)n + k - 1$. For example, 21, 2133, and 19521 are 2-hyperperfect and 325 is 3-hyperperfect. They conjecture that there are $k$-hyperperfect numbers for every $k$.

Graham asks if $s(n) = \lfloor n/2 \rfloor$ implies that $n$ is 2 or a power of 3.

Erdös lets $f(n)$ be the smallest integer for which $n = \sum_{i=1}^{k} d_i$ for some $k$, where $1 = d_1 < d_2 < \cdots < d_l = f(n)$ is the increasing sequence of divisors of $f(n)$. Is $f(n) = o(n)$? Or is this true only for almost all $n$?

| $n$ | 1 | 2 | 3 | 4 | 5 | 6 | 7 | 8 | 9 | 10 | 11 | 12 | 13 | 14 | 15 | 16 | 17 | 18 | 19 | 20 | 21 | 22 | 23 | 24 | 25 | 26 | 27 | 28 |
|---|---|---|---|---|---|---|---|---|---|---|---|---|---|---|---|---|---|---|---|---|---|---|---|---|---|---|---|---|
| $f(n)$ | 1 | – | 2 | 3 | – | 5 | 4 | 7 | 15 | 12 | 21 | 6 | 9 | 13 | 8 | 12 | 30 | 10 | 42 | 19 | 18 | 20 | 57 | 14 | 36 | 46 | 30 | 12 |


H. Abbott, C. E. Aull, Ezra Brown, and D. Suryanarayana, Quasiperfect numbers, *Acta Arith.* **22** (1973) 439–447; *MR* 47 #4915. Corrections, *ibid.* **29** (1976) 427–428.

M. M. Artuhov, On the problem of odd $h$-fold perfect numbers, *Acta Arith.* **23** (1973) 249–255.

P. T. Bateman, P. Erdös, C. Pomerance, and E. G. Straus, in *Proc. Grosswald Conf.*, Springer-Verlag, New York, 1980.

S. J. Benkoski, Problem E. 2308, *Amer. Math. Monthly* **79** (1972) 774.

S. J. Benkoski and P. Erdös, On weird and pseudoperfect numbers, *Math. Comput.* **28** (1974) 617–623; *MR* **50** #228 (Corrigendum, S. Kravitz, *ibid.* **29** (1975) 673).

Alan L. Brown, Multiperfect numbers, *Scripta Math.* **20** (1954) 103–106; *MR* **16**, 12.

Paolo Catteneo, Sui numeri quasiperfetti, *Boll. Un. Mat. Ital.* (3) **6** (1951) 59–62; *Zbl.* 42, 268.

Graeme L. Cohen, The non-existence of quasiperfect numbers of certain forms (to appear).

G. L. Cohen and M. D. Hendy, On odd multiperfect numbers, *Math. Chronicle* **9** (1980).

J. T. Cross, A note on almost perfect numbers, *Math. Mag.* **47** (1974) 230–231.

P. Erdös, Problems in number theory and combinatorics, *Congressus Numerantium XVIII, Proc. 6th Conf. Numerical Math. Manitoba*, 1976, 35–58 (esp. pp. 53–54); *MR* **80e**:10005.

Benito Franqui and Mariano Garcia, Some new multiply perfect numbers, *Amer. Math. Monthly*, **60** (1953) 459–462.

Benito Franqui and Mariano Garcia, 57 new multiply perfect numbers, *Scripta Math.* **29** (1954) 169–171 (1955).

Mariano Garcia, A generalization of multiply perfect numbers, *Scripta Math.* **19** (1953) 209–210.

Mariano Garcia, On numbers with integral harmonic mean, *Amer. Math. Monthly*, **61** (1954) 89–96.

P. Hagis and G. L. Cohen, Some results concerning quasiperfect numbers, *J. Austral. Math. Soc.*, (to appear).

B. Hornfeck and E. Wirsing, Über die Häufigkeit vollkommener Zahlen, *Math. Ann.* **133** (1957) 431–438; *MR* **19**, 837. See also *ibid.* **137** (1959) 316–318; *MR* **21** #3389.

R. P. Jerrard and Nicholas Temperley, Almost perfect numbers, *Math. Mag.* **46** (1973) 84–87.
H.-J. Kanold, Über das harmonische Mittel der Teiler einer natürlichen Zahl, *Math. Ann.* **133** (1957) 371–374.
David G. Kendall, The scale of perfection, *J. Appl. Probability* **19A** (P.A.P. Moran birthday volume, 1982.) (to appear)
Masao Kishore, Odd almost perfect numbers, *Notices Amer. Math. Soc.* **22** (1975) A-380.
Masao Kishore, Quasiperfect numbers are divisible by at least six distinct prime factors, *Notices Amer. Math. Soc.* **22** (1975) A-441.
Masao Kishore, Odd integers $N$ with 5 distinct prime factors for which $2 - 10^{-12} < \sigma(N)/N < 2 + 10^{-12}$, *Math. Comput.* **32** (1978) 303–309.
M. S. Klamkin, Problem E.1445*, *Amer. Math. Monthly* **67** (1960) 1028. See also *ibid.* **82** (1975) 73.
Sydney Kravitz, A search for large weird numbers, *J. Recreational Math.* **9** (1976–77) 82–85.
A. Mąkowski, Remarques sur les fonctions $\theta(n)$, $\phi(n)$ et $\sigma(n)$, *Mathesis* **69** (1960) 302–303.
A. Mąkowski, Some equations involving the sum of divisors, *Elem. Math.* **34** (1979) 82; *MR* **81b**:10004.
D. Minoli, Structural issues for hyperperfect numbers, *Fibonacci Quart.*, (to appear).
D. Minoli, Issues in non-linear hyperperfect numbers, *Math. Comput.* **34** (1980) 639–645.
Daniel Minoli and Robert Bear, Hyperperfect numbers, *Pi Mu Epsilon J.* **6** #3 (1974–75) 153–157.
Oystein Ore, On the averages of the divisors of a number, *Amer. Math. Monthly* **55** (1948) 615–619.
Seppo Pajunen, On primitive weird numbers, *A collection of manuscripts related to the Fibonacci sequence*, 18th anniv. vol., Fibonacci Assoc., 162–166.
Carl Pomerance, On a problem of Ore: harmonic numbers (unpublished typescript).
Carl Pomerance, On multiply perfect numbers with a special property, *Pacific J. Math.* **57** (1975) 511–517.
Carl Pomerance, On the congruences $\sigma(n) \equiv a \pmod{n}$ and $n \equiv a \pmod{\phi(n)}$, *Acta Arith.* **26** (1975) 265–272.
Paul Poulet, *La Chasse aux Nombres*, Fascicule I, Bruxelles, 1929, 9–27.
Problem B-6, William Lowell Putnam Mathematical Competition, 1976:12:04.
Herman J. J. te Riele, Hyperperfect numbers with three different prime factors, *Math. Comput.* **36** (1981) 297–298.
Neville Robbins, A class of solutions of the equation $\sigma(n) = 2n + t$, *Fibonacci Quart.* **18** (1980) 137–147 (misprints in solutions for $t = 31, 84, 86$).
H. N. Shapiro, Note on a theorem of Dickson, *Bull. Amer. Math. Soc.* **55** (1949) 450–452.
H. N. Shapiro, On primitive abundant numbers, *Comm. Pure Appl. Math.* **21** (1968) 111–118.
W. Sierpiński, Sur les nombres pseudoparfaits, *Mat. Vesnik* **2** (17) (1965) 212–213; *MR* **33** #7296.
W. Sierpiński, *Number Theory* (in Polish), 1959, p. 257.
D. Suryanarayana, Quasi-perfect numbers II, *Bull. Calcutta Math. Soc.* **69** (1977) 421–426; *MR* **80m**:10003.
Andreas and Eleni Zachariou, Perfect, semi-perfect and Ore numbers, *Bull. Soc. Math. Grèce* (N.S.) **13** (1972) 12–22; *MR* **50** #12905.

**B3.** If $d$ divides $n$ and $(d, n/d) = 1$, call $d$ a **unitary divisor** of $n$. A number $n$ which is the sum of its unitary divisors, apart from $n$, is a **unitary perfect number**. There are no odd unitary perfect numbers, and Subbarao conjectures that there are only a finite number of even ones. He, Carlitz, and Erdös each offer \$10.00 for settling this question and Subbarao offers 10¢ for each new example. If $n = 2^a m$, where $m$ is odd and has $r$ distinct prime divisors, then Subbarao and others have shown that, apart from $2 \times 3$, $2^2 \times 3 \times 5$, $2 \times 3^2 \times 5$, and $2^6 \times 3 \times 5 \times 7 \times 13$, there are no unitary perfect numbers with $a \leq 10$, or with $r \leq 6$. Wall has found the unitary perfect number

$$2^{18} \times 3 \times 5^4 \times 7 \times 11 \times 13 \times 19 \times 37 \times 79 \times 109 \times 157 \times 313$$

and shown that it is the fifth such. Frey has shown that if $N = 2^m p_1^{a_1} \cdots p_r^{a_r}$ is unitary perfect with $(N, 3) = 1$, then $m > 144$, $r > 144$ and $N > 10^{440}$.

H. A. M. Frey, Über unitär perfekte Zahlen, *Elem. Math.* **33** (1978) 95–96; *MR* **81a**: 10007.
M. V. Subbarao, Are there an infinity of unitary perfect numbers?, *Amer. Math. Monthly* **77** (1970) 389–390.
M. V. Subbarao and D. Suryanarayana, Sums of the divisor and unitary divisor functions, *J. reine angew. Math.* **302** (1978) 1–15; *MR* **80d**:10069.
M. V. Subbarao and L. J. Warren, Unitary perfect numbers, *Canad. Math. Bull.* **9** (1966) 147–153; *MR* **33** #3994.
M. V. Subbarao, T. J. Cook, R. S. Newberry, and J. M. Weber, On unitary perfect numbers, *Delta* **3** #1 (Spring 1972) 22–26.
Charles R. Wall, The fifth unitary perfect number, *Canad. Math. Bull.* **18** (1975) 115–122. See also *Notices Amer. Math. Soc.* **16** (1969) 825.

**B4.** Unequal numbers $m$, $n$ are called **amicable** if each is the sum of the aliquot parts of the other, i.e., $\sigma(m) = \sigma(n) = m + n$. Over a thousand such pairs are known. The smaller member, 220, of the smallest pair, occurs in *Genesis*, xxxii, 14, and amicable numbers intrigued the Greeks and Arabs and many others since. For their history, see the articles of Lee and Madachy.

It is not known if there are infinitely many, but it is believed that there are. In fact Erdös conjectures that the number, $A(x)$, of such pairs with $m < n < x$ is at least $cx^{1-\varepsilon}$. He improved a result of Kanold to show that $A(x) = o(x)$ and his method can be used to obtain $A(x) \leq cx/\ln\ln\ln x$, while Pomerance has obtained the further improvement $A(x) \leq x \exp\{-c(\ln\ln\ln x \ln\ln\ln\ln x)^{1/2}\}$. Erdös conjectures that $A(x) = o(x/(\ln x)^k)$ for every $k$. As we go to press Pomerance confirms this conjecture with the "dramatically stronger result"

$$A(x) \leq x \exp\{-(\ln x)^{1/3}\}.$$

This also implies that the sum of the reciprocals of the amicable numbers is finite, a fact not earlier known. He also notes that his proof can be modified to give the slightly stronger result

$$A(x) \ll x \exp\{-c(\ln x \ln\ln x)^{1/3}\}.$$

It is not known if an amicable pair exists with $m$ and $n$ of opposite parity, or with $(m, n) = 1$. Bratley and McKay conjecture that both members of all odd amicable pairs are divisible by 3.

Some very large amicable pairs, with 32, 40, 81, and 152 decimal digits, discovered by te Riele, are mentioned by Kaplansky under "Mathematics" in the 1975 *Encyclopedia Brittanica Yearbook*. The largest previously known had 25 decimal digits.

J. Alanen, O. Ore, and J. G. Stemple, Systematic computations on amicable numbers, *Math. Comput.* **21** (1967) 242–245; *MR* **36** #5058.
M. M. Artuhov, On some problems in the theory of amicable numbers (Russian), *Acta Arith.* **27** (1975) 281–291.
W. Borho, On Thabit ibn Kurrah's formula for amicable numbers, *Math. Comput.* **26** (1972) 571–578.
W. Borho, Befreundete Zahlen mit gegebener Primteileranzahl, *Math. Ann.* **209** (1974) 183–193.
W. Borho, Eine Schranke für befreundete Zahlen mit gegebener Teileranzahl, *Math. Nachr.* **63** (1974) 297–301.
W. Borho, Some large primes and amicable numbers, *Math. Comput.* **36** (1981) 303–304.
P. Bratley and J. McKay, More amicable numbers, *Math. Comput.* **22** (1968) 677–678; *MR* **37** #1299.
P. Bratley, F. Lunnon, and J. McKay, Amicable numbers and their distribution, *Math. Comput.* **24** (1970) 431–432.
B. H. Brown, A new pair of amicable numbers, *Amer. Math. Monthly*, **46** (1939) 345.
Patrick Costello, Four new amicable pairs, *Notices Amer. Math. Soc.* **21** (1974) A-483.
Patrick Costello, Amicable pairs of Euler's first form, *Notices Amer. Math. Soc.* **22** (1975) A-440.
P. Erdös, On amicable numbers, *Publ. Math. Debrecen* **4** (1955) 108–111; *MR* **16**, 998.
P. Erdös and G. J. Rieger, Ein Nachtrag über befreundete Zahlen, *J. reine angew. Math.* **273** (1975) 220.
E. B. Escott, Amicable numbers, *Scripta Math.* **12** (1946) 61–72; *MR* **8**, 135.
M. Garcia, New amicable pairs, *Scripta Math.* **23** (1957) 167–171; *MR* **20** #5158.
A. A. Gioia and A. M. Vaidya, Amicable numbers with opposite parity, *Amer. Math. Monthly* **74** (1967) 969–973; correction **75** (1968) 386; *MR* **36** #3711, **37** #1306.
P. Hagis, On relatively prime odd amicable numbers, *Math. Comput.* **23** (1969) 539–543; *MR* **40** #85.
P. Hagis, Lower bounds for relatively prime amicable numbers of opposite parity, *ibid.* **24** (1970) 963–968.
P. Hagis, Relatively prime amicable numbers of opposite parity, *Math. Mag.* **43** (1970) 14–20.
H.-J. Kanold, Über die Dichten der Mengen der vollkommenen und der befreundeten Zahlen, *Math. Z.* **61** (1954) 180–185; *MR* **16**, 337.
H.-J. Kanold, Über befreundete Zahlen I, *Math. Nachr.* **9** (1953) 243–248; II, *ibid.* **10** (1953) 99–111; *MR* **15**, 506.
H.-J. Kanold, Über befreundete Zahlen III, *J. reine angew. Math.* **234** (1969) 207–215; *MR* **39** #122.
E. J. Lee, Amicable numbers and the bilinear diophantine equation, *Math. Comput.* **22** (1968) 181–187; *MR* **37** #142.
E. J. Lee, On divisibility by nine of the sums of even amicable pairs, *Math. Comput.* **23** (1969) 545–548; *MR* **40** #1328.
E. J. Lee and J. S. Madachy, The history and discovery of amicable numbers, part 1, *J. Recreational Math.* **5** (1972) 77–93; part 2, *ibid.* 153–173; part 3, *ibid.* 231–249.
O. Ore, *Number Theory and its History*, McGraw-Hill, New York, 1948, p. 89.
Carl Pomerance, On the distribution of amicable numbers, *J. reine angew. Math.* **293/294** (1977) 217–222; II *ibid.* (1981).
P. Poulet, 43 new couples of amicable numbers, *Scripta Math.* **14** (1948) 77.
H. J. J. te Riele, Four large amicable pairs, *Math. Comput.* **28** (1974) 309–312.

**B5.** Garcia has called a pair of numbers $(m, n)$, $m < n$, *quasi-amicable* if

$$\sigma(m) = \sigma(n) = m + n + 1.$$

For example, (48, 75), (140, 195), (1575, 1648), (1050, 1925), and (2024, 2295). Rufus Isaacs, noting that each of $m$ and $n$ is the sum of the *proper* divisors of the other (i.e., omitting 1 as well as the number itself) has much more appropriately named them **betrothed numbers**.

Hagis and Lord have found all 46 pairs with $m < 10^7$. All of them are of opposite parity. No pairs are known with $m$, $n$ having the same parity. If there are such, then $m > 10^{10}$. If $(m, n) = 1$, then $mn$ contains at least four distinct prime factors, and if $mn$ is odd, then $mn$ contains at least 21 distinct prime factors.

Beck and Najar call such pairs *reduced* amicable pairs, and call numbers $m$, $n$ such that

$$\sigma(m = \sigma(n) = m + n - 1$$

*augmented* amicable pairs. They found 11 augmented amicable pairs. They found no reduced or augmented *unitary* amicable or sociable numbers (see B8) with $n < 10^5$.


Walter E. Beck and Rudolph M. Najar, More reduced amicable pairs, *Fibonacci Quart.* **15** (1977) 331–332; *Zbl.* 389.10004.

Walter E. Beck and Rudolph M. Najar, Fixed points of certain arithmetic functions, *Fibonacci Quart.* **15** (1977) 337–342; *Zbl.* 389.10005.

Peter Hagis and Graham Lord, Quasi-amicable numbers, *Math. Comput.* **31** (1977) 608–611; *MR* **55** #7902; *Zbl.* 355.10010.

M. Lal and A. Forbes, A note on Chowla's function, *Math. Comput.* **25** (1971) 923–925; *MR* **45** #6737; *Zbl.* 245.10004.


**B6. Aliquot sequences.** Since some numbers are abundant and some deficient, it is natural to ask what happens when you iterate the sum of divisors function, and produce an aliquot sequence, $\{s^k(n)\}$, $k = 0, 1, 2, \ldots$ . Catalan and Dickson conjectured that all such sequences were bounded, but we now have heuristic arguments and experimental evidence that some sequences, perhaps almost all of those with $n$ even, go to infinity. The smallest $n$ for which there was ever doubt was 138, but D. H. Lehmer eventually showed that after reaching a maximum

$$s^{117}(138) = 179931\ 895322 = 2 \times 61 \times 929 \times 1587569$$

the sequence terminated at $s^{177}(138) = 1$. The next value for which there continues to be real doubt is 276. A good deal of computation by Lehmer, subsequently assisted by Godwin, Selfridge, Wunderlich, and others, have established that

$$s^{469}(276) = 149\ 384846\ 598254\ 844243\ 905695\ 992651\ 412919\ 855640.$$

H. W. Lenstra has proved that it is possible to construct arbitrarily long monotonic increasing aliquot sequences.

Jack Alanen, Empirical study of aliquot series, *Math. Rep.* **133**, Stichting Math. Centrum, Amsterdam, 1972; reviewed *Math. Comput.* **28** (1974) 878–880.
E. Catalan, *Bull. Soc. Math. France* **16** (1887–88) 128–129.
John S. Devitt, Aliquot Sequences, MSc thesis, The Univ. of Calgary, 1976; see *Math. Comput.* **32** (1978) 942–943.
J. S. Devitt, R. K. Guy, and J. L. Selfridge, Third report on aliquot sequences, *Congressus Numeratium* XVIII, Proc. 6th Manitoba Conf. Numerical Math. 1976, 177–204; *MR* **80d**:10001.
L. E. Dickson, Theorems and tables on the sum of the divisors of a number, *Quart. J. Math.* **44** (1913) 264–296.
Paul Erdös, On asymptotic properties of aliquot sequences, *Math. Comput.* **30** (1976) 641–645.
Richard K. Guy, Aliquot sequences, in *Number Theory and Algebra*, Academic Press, 1977, 111–118; *MR* **57** #223; *Zbl.* 367.10007.
Richard K. Guy and J. L. Selfridge, Interim report on aliquot series, *Congressus Numerantium* V, Proc. Conf. Numerical Math. Winnipeg, 1971, 557–580; *MR* **49** #194; *Zbl.* 266.10006.
Richard K. Guy and J. L. Selfridge, Combined report on aliquot sequences, The Univ. of Calgary Math. Res. Report No. 225, May, 1974.
Richard K. Guy and J. L. Selfridge, What drives an aliquot sequence? *Math. Comput.* **29** (1975) 101–107; *MR* **52** #5542; *Zbl.* 296.10007. Corrigendum, *ibid.* **34** (1980).
Richard K. Guy and M. R. Williams, Aliquot sequences near $10^{12}$, *Congressus Numerantium* XII, Proc. 4th Conf. Numerical Math. Winnipeg, 1974, 387–406; *MR* **52** #242; *Zbl.* 359.10007.
Richard K. Guy, D. H. Lehmer, J. L. Selfridge, and M. C. Wunderlich, Second report on aliquot sequences, *Congressus Numerantium* IX, Proc. 3rd Conf. Numerical Math. Winnipeg, 1973, 357–368; *MR* **50** #4455; *Zbl.* 325.10007.
G. Aaron Paxson, Aliquot sequences (preliminary report), *Amer. Math. Monthly*, **63** (1956) 614. See also *Math. Comput.* **26** (1972) 807–809.
P. Poulet, La chasse aux nombres, Fascicule I, Bruxelles, 1929.
H. J. J. te Riele, A note on the Catalan–Dickson conjecture, *Math. Comput.* **27** (1973) 189–192; *MR* **48** #3869; *Zbl.* 255.10008.

**B7. Aliquot cycles** or **sociable numbers**. Poulet discovered two cycles of numbers, showing that $s^k(n)$ can have the periods 5 and 28, in addition to 1 and 2. For $k \equiv 0, 1, 2, 3, 4 \pmod 5$, $s^k(12496)$ takes the values

$$12496 = 2^4 \times 11 \times 71, \quad 14288 = 2^4 \times 19 \times 47, \quad 15472 = 2^4 \times 967,$$
$$14536 = 2^3 \times 23 \times 79, \quad 14264 = 2^3 \times 1783.$$

For $k \equiv 0, 1, \ldots, 27 \pmod{28}$, $s^k(14316)$ takes the values

| | | | | | | |
|---|---|---|---|---|---|---|
| 14316 | 19116 | 31704 | 47616 | 83328 | 177792 | 295488 |
| 629072 | 589786 | 294896 | 358336 | 418904 | 366556 | 274924 |
| 275444 | 243760 | 376736 | 381028 | 285778 | 152990 | 122410 |
| 97946 | 48976 | 45946 | 22976 | 22744 | 19916 | 17716 |

After a gap of over 50 years, and the advent of high-speed computing, Henri Cohen discovered nine cycles of period 4, and Borho, David, and Root discovered others. The smallest members of these 4-cycles are

| | | | | |
|---|---|---|---|---|
| 1264460 | 2115324 | 2784580 | 4938136 | 7169104 |
| 18048976 | 18656380 | 28158165 | 46722700 | 81128632 |
| | 174277820 | 209524210 | 330003580 | 498215416 |

It has been conjectured that there are no 3-cycles.

W. Borho, Über die Fixpunkte der $k$-fach iterierten Teilersummenfunktion, *Mitt. Math. Gesellsch. Hamburg* **4** (1969) 35–38; *MR* **40** #7189.
H. Cohen, On amicable and sociable numbers, *Math. Comput.* **24** (1970) 423–429; *MR* **42** #5887.
Richard David, letter to D. H. Lehmer, 1972:02:25.
P. Poulet, Question 4865, *L'Intermédiaire des math.* **25** (1918) 100–101.
S. C. Root, in M. Beeler, R. W. Gosper and R. Schroeppel, M.I.T. Artificial Intelligence Memo 239, 1972:02:29.

**B8.** The ideas of aliquot sequence and aliquot cycle can be adapted to the case where only the *unitary* divisors are summed, leading to **unitary aliquot sequences** and **unitary sociable numbers**. We use $\sigma^*(n)$ and $s^*(n)$ for the analogs of $\sigma(n)$ and $s(n)$ when just the unitary divisors are summed (compare B3).

Are there unbounded unitary aliquot sequences? Here the balance is more delicate than in the ordinary aliquot sequence case. The only sequences which deserve serious consideration are the odd multiples of 6, which is a unitary perfect number as well as an ordinary one. Now the sequences tend to increase if $3 \| n$, but decrease when a higher power of 3 is present, and it is a moot point as to which situation will dominate. Once a term of a sequence is $6m$, with $m$ odd, then $\sigma^*(6m)$ is an even multiple of 6, making $s^*(6m)$ an odd multiple of 6 again, except in the extremely rare case that $m$ is an odd power of 3.

te Riele pursued all unitary aliquot sequences for $n < 10^5$. The only one which did not terminate or become periodic was 89610. Later calculations showed that this reached a maximum,

$$645\ 856907\ 610421\ 353834 = 2 \times 3^2 \times 13 \times 19 \times 73 \times 653 \times 3047409443791$$

at its 568th term, and terminated at its 1129th.

One can hardly expect typical behavior until the expected number of prime factors is large. Since this number is $\ln \ln n$, such sequences are well beyond computer range. Of 80 sequences examined near $10^{12}$, all have terminated or become periodic. One sequence exceeded $10^{23}$.

Unitary amicable pairs and unitary sociable numbers may occur rather more frequently than their ordinary counterparts. Lal, Tiller, and Summers found cycles of periods 1, 2, 3, 4, 5, 6, 14, 25, 39, and 65. Examples of unitary amicable pairs are (56430, 64530) and (1080150, 1291050), while (30, 42, 54) is a 3-cycle and (1482, 1878, 1890, 2142, 2178) is a 5-cycle.

Erdös, looking for a number-theoretic function whose iterates might be bounded, suggested defining $w(n) = n \sum 1/p_i^{\alpha_i}$ where $n = \prod p_i^{\alpha_i}$, and $w^k(n) = w(w^{k-1}(n))$. Note that $(w(n), n) = 1$. Can it be proved that $w^k(n)$, $k = 1, 2, \ldots$, is bounded? Is $|\{w(n): 1 \le n \le x\}| = o(x)$?

Erdös and Selfridge called $n$ a **barrier** for a number-theoretic function $f(m)$ if, for all $m < n$, $m + f(m) \le n$. Euler's $\phi$-function (see B36) and $\sigma(m)$ increase too fast to have barriers, but does $\omega(m)$, the number of distinct prime factors of $m$, have infinitely many barriers? The numbers 2, 3, 4, 5, 6, 8, 9, 10, 12, 14, 17, 18, 20, 24, 26, 28, 30, . . . are barriers for $\omega(m)$. If $\Omega(m)$ is the number of prime factors of $m$, not necessarily distinct, does $\Omega(m)$ have infinitely many barriers? Selfridge observes that 99840 is the largest barrier for $\Omega(m)$ that is $< 10^5$. The number of divisors of $m$, $d(m)$, does not have barriers, because $\max\{d(n-1) + n - 1, d(n-2) + n - 2\} \ge n + 2$. Does

$$\max_{m<n} (m + d(m)) = n + 2$$

have infinitely many solutions? It is very doubtful. One solution is $n = 24$; the next larger is probably beyond computer range.


Paul Erdös, A mélange of simply posed conjectures with frustratingly elusive solutions, *Math. Mag.* **52** (1979) 67–70.

P. Erdös, Problems and results in number theory and graph theory, *Congressus Numerantium* XXVII, Proc. 9th Manitoba Conf. Numerical Math. Comput. 1979, 3–21.

Richard K. Guy and Marvin C. Wunderlich, Computing unitary aliquot sequences—a preliminary report, *Congressus Numerantium* XXVII, Proc. 9th Manitoba Conf. Numerical Math. Comput. 1979, 257–270.

P. Hagis, Unitary amicable numbers, *Math. Comput.* **25** (1971) 915–918; *MR* **45** #8599; *Zbl.* 232.10004.

Peter Hagis, Unitary hyperperfect numbers, *Math. Comput.* **36** (1981) 299–301.

M. Lal, G. Tiller and T. Summers, Unitary sociable numbers, *Congressus Numerantium* VII, Proc. 2nd Conf. Numerical Math. Winnipeg, 1972, 211–216; *MR* **50** #4471; *Zbl.* 309.10005.

H. J. J. te Riele, *Unitary Aliquot Sequences*, *MR* 139/72, Mathematisch Centrum, Amsterdam, 1972; reviewed *Math. Comput.* **32** (1978) 944–945; *Zbl.* 251.10008.

H. J. J. te Riele, *Further Results on Unitary Aliquot Sequences*, NW2/73, Mathematisch Centrum, Amsterdam, 1973; reviewed *Math. Comput.* **32** (1978) 945.

H. J. J. te Riele, *A Theoretical and Computational Study of Generalized Aliquot Sequences*, MCT74, Mathematisch Centrum, Amsterdam, 1976; reviewed *Math. Comput.* **32** (1978) 945–946; *MR* **58** #27716.

C. R. Wall, Topics related to the sum of unitary divisors of an integer, PhD thesis, Univ. of Tennessee, 1970.


**B9.** Suryanarayana defines **superperfect numbers** $n$ by $\sigma^2(n) = 2n$, i.e., $\sigma(\sigma(n)) = 2n$. He and Kanold show that the even ones are just the numbers $2^{p-1}$, where $2^p - 1$ is a Mersenne prime. Are there any odd superperfect numbers? If so, Kanold shows that they are perfect squares, and Dandapat et al that $n$ or $\sigma(n)$ is divisible by at least three distinct primes.

More generally, Bode defines $m$-**perfect numbers** as numbers $n$ for which $\sigma^m(n) = 2n$, and shows that for $m \geq 3$ there are no even $m$-perfect numbers. He also shows that (for $m = 2$) there is no odd superperfect number $< 10^{10}$. Hunsucker and Pomerance have raised this bound to $7 \times 10^{24}$ and have unpublished results on the numbers of distinct prime factors of $n$ and $\sigma(n)$ if $n$ is superperfect.

If $\sigma^2(n) = 2n + 1$, it would be consistent with earlier terminology to call $n$ quasi-superperfect! The Mersenne primes are such. Are there others? Are there "almost superperfect numbers," for which $\sigma^2(n) = 2n - 1$?

Erdös asks if $(\sigma^k(n))^{1/k}$ has a limit as $k \to \infty$. He conjectures that it is infinite for each $n > 1$.

Schinzel asks if lim inf $\sigma^k(n)/n < \infty$ for each $k$, as $n \to \infty$, and observes that it follows for $k = 2$ from a deep theorem of Rényi. Mąkowski and Schinzel give an elementary proof for $k = 2$ that the limit is 1.

Dieter Bode, Über eine Verallgemeinerung der Vollkommenen Zahlen, Dissertation, Braunschweig, 1971.

P. Erdös, Some remarks on the iterates of the $\phi$ and $\sigma$ functions, *Colloq. Math.* **17** (1967) 195–202.

J. L. Hunsucker and C. Pomerance, There are no odd superperfect numbers less than $7 \times 10^{24}$, *Indian J. Math.* **17** (1975) 107–120.

H.-J. Kanold, Über "Super perfect numbers," *Elem. Math.* **24** (1969) 61–62; *MR* **39** #5463.

Graham Lord, Even perfect and superperfect numbers, *Elem. Math.* **30** (1975) 87–88.

A. Mąkowski and A. Schinzel, On the functions $\phi(n)$ and $\sigma(n)$, *Colloq. Math.* **13** (1964–65) 95–99.

A. Schinzel, Ungelöste Probleme Nr. 30, *Elem. Math.* **14** (1959) 60–61.

D. Suryanarayana, Super perfect numbers, *Elem. Math.* **24** (1969) 16–17; *MR* **39** #5706.

D. Suryanarayana, There is no odd superperfect number of the form $p^{2\alpha}$, *Elem. Math.* **28** (1973) 148–150.

**B10.** Erdös has proved that there are infinitely many $n$ such that $s(x) = n$ has no solution. Alanen calls such $n$ **untouchable**. In fact Erdös shows that the untouchable numbers have positive lower density. Here are the untouchable numbers less than 1000:

2 5 52 88 96 120 124 146 162 178 188 206 210 216 238 246
248 262 268 276 288 290 292 304 307 322 324 326 336 342 372 406
408 426 430 448 472 474 498 516 518 520 530 540 552 556 562 576
584 612 624 626 628 658 668 670 714 718 726 732 738 748 750 756
766 768 782 784 792 802 804 818 836 848 852 872 892 894 896 898
902 916 926 936 964 966 976 982 996

In view of the plausibility of the Goldbach conjecture, it seems likely that 5 is the only odd untouchable number since if $2n + 1 = p + q + 1$ with $p$ and $q$ prime, then $s(pq) = 2n + 1$. Can this be proved independently? Are there arbitrarily long sequences of consecutive even numbers which are untouchable? How large can the gaps between untouchable numbers be?

P. Erdös, Über die Zahlen der Form $\sigma(n) - n$ und $n - \phi(n)$, *Elem. Math.* **28** (1973) 83–86.

Paul Erdös, Some unconventional problems in number theory, *Astérisque* **61** (1979) 73–82; *Zbl.* 399.10001; *MR* **81h**: 10001.

**B11.** Leo Moser has observed that while $n\phi(n)$ determines $n$ uniquely, $n\sigma(n)$ does not ($\phi(n)$ is Euler's totient function; see B36). For example $m\sigma(m) = n\sigma(n)$ for $m = 12$, $n = 14$. The multiplicativity of $\sigma(n)$ now ensures an infinity of solutions, $m = 12q$, $n = 14q$, where $(q, 42) = 1$. So Moser asked if there is an infinity of *primitive* solutions, in the sense that $(m^*, n^*)$ is *not* a solution for any $m^* = m/d$, $n^* = n/d$, $d > 1$. The example we've given is the least of the set $m = 2^{p-1}(2^q - 1)$, $n = 2^{q-1}(2^p - 1)$, where $2^p - 1$, $2^q - 1$ are distinct Mersenne primes, so that only a finite number of such solutions is known. Another set of solutions is $m = 2^7 \times 3^2 \times 5^2 \times (2^p - 1)$, $n = 2^{p-1} \times 5^3 \times 17 \times 31$, where $2^p - 1$ is a Mersenne prime other than 3 or 31; also $p = 5$ gives a primitive solution on deletion of the common factor 31. There are other solutions, such as $m = 2^4 \times 3 \times 5^3 \times 7$, $n = 2^{11} \times 5^2$ and $m = 2^9 \times 5$, $n = 2^3 \times 11 \times 31$. An example with $(m, n) = 1$ is $m = 2^5 \times 5$, $n = 3^3 \times 7$.

Erdös observes that if $n$ is squarefree, then integers of the form $n\sigma(n)$ are distinct. He can also prove that the number of solutions of $m\sigma(m) = n\sigma(n)$ with $m < n < x$ is $cx + o(x)$. Are there *three* distinct numbers $l$, $m$, $n$ such that $l\sigma(l) = m\sigma(m) = n\sigma(n)$? Is there an infinity of primitive solutions of the equation $\sigma(a)/a = \sigma(b)/b$? Without restricting the solutions to being primitive, Erdös can show that their number with $a < b < x$ is $cx + o(x)$; with the restriction $(a, b) = 1$, no solution is known at all.

Erdös believes that the number of solutions of $x\sigma(x) = n$ is less than $n^{\varepsilon/\ln\ln n}$ for every $\varepsilon > 0$, and says that the number may be less than $(\ln n)^c$.

P. Erdös, Remarks on number theory II: some problems on the $\sigma$ function, *Acta Arith.* **5** (1959) 171–177; *MR* **21** #6348.

**B12.** Analogous question may be asked with $\sigma_k(n)$ in place of $\sigma(n)$, where $\sigma_k(n)$ is the sum of the $k$th powers of the divisors of $n$: are there distinct numbers $m$, $n$ such that $m\sigma_2(m) = n\sigma_2(n)$? For the case $k = 0$ we have $md(m) = nd(n)$ for $(m, n) = (18, 27)$, $(24, 32)$, $(56, 64)$, and $(192, 224)$. The last pair can be supplemented by 168 to give three distinct numbers such that $ld(l) = md(m) = nd(n)$. There is an infinity of primitive solutions $(m, n)$; for example

$$m = 2^{qt-1}p, \qquad n = 2^{pt \cdot 2^{tu}-1}q$$

where $p$ and $q = u + p \cdot 2^{tu}$ are primes. Many other solutions can be constructed; for example $(2^{70}, 2^{63} \times 71)$, $(3^{19}, 3^{17} \times 5)$ and $(5^{51}, 5^{49} \times 13)$.

**B13.** Sierpiński has asked if $\sigma(n) = \sigma(n + 1)$ infinitely often. Hunsucker et al extended the tabulations of Makowski and of Mientka and Vogt, and have found the 113 solutions

$$14, 206, 957, 1334, 1364, 1634, 2685, 2974, 4364, \ldots$$

less than $10^7$. They also obtain statistics concerning the equation $\sigma(n) = \sigma(n + l)$, of which Mientka and Vogt had asked: for what $l$ (if any) is there an infinity of solutions? They found many solutions if $l$ was a factorial, but only two solutions for $l = 15$ and $l = 69$. They also ask whether, for each $l$ and $m$, there is an $n$ such that $\sigma(n) + m = \sigma(n + l)$.

One can ask corresponding questions for $\sigma_k(n)$, the sum of the $k$th powers of the divisors of $n$ (for $k=0$, see B15). The only solution of $\sigma_2(n)=\sigma_2(n+1)$ is $n = 6$, since $\sigma_2(2n) > \sigma_2(2n + 1)$ for $n > 7$. Note that $\sigma_2(24) = \sigma_2(26)$; Erdös doubts that $\sigma_2(n) = \sigma_2(n + 2)$ has infinitely many solutions, and thinks that $\sigma_3(n) = \sigma_3(n + 2)$ has no solutions at all.

Richard K. Guy and Daniel Shanks, A constructed solution of $\sigma(n) = \sigma(n + 1)$, *Fibonacci Quart.* **12** (1974) 299; *MR* **50** #219; *Zbl.* 287.10004.

John L. Hunsucker, Jack Nebb, and Robert E. Stearns, Computational results concerning some equations involving $\sigma(n)$, *Math. Student* **41** (1973) 285–289.

A. Mąkowski, On some equations involving functions $\varphi(n)$ and $\sigma(n)$, *Amer. Math. Monthly* **67** (1960) 668–70· correction, *ibid.* **68** (1971) 650.

W.E. Mientka and R. L. Vogt, Computational results relating to problems concerning $\sigma(n)$, *Mat. Vesnik* **7** (1970) 35–36.

**B14.** Is $\sum_{n=1}^{\infty} (\sigma_k(n)/n!)$ irrational? It is for $k = 1$ and 2.

P. Erdös and M. Kac, Problem 4518, *Amer. Math. Monthly* **60** (1953) 47. Solution, R. Breusch, **61** (1954) 264–265.

**B15.** Max Rumney (*Eureka*, **26** (1963) 12) asked if the equation $\sigma(q) + \sigma(r) = \sigma(q + r)$ has infinitely many solutions which are primitive in a sense similar to that used in B11. If $q + r$ is prime, the only solution is $(q, r) = (1, 2)$. If $q + r = p^2$, where $p$ is prime, then one of $q$ and $r$, say $q$, is prime, and $r = 2^n k^2$ where $n \geq 1$ and $k$ is odd. If $k = 1$, there is a solution if $p = 2^n - 1$ is a Mersenne prime and $q = p^2 - 2^n$ is prime; this is so for $n = 2, 3, 5, 7, 13$, and 19. For $k = 3$ there are no solutions, and none for $k=5$ with $n<189$. For $k=7$, $n=1$ and 3 give $(q, r, q+r)=(5231, 2 \times 7^2, 73^2)$ and $(213977, 2^3 \times 7^2, 463^2)$. Other solutions are $(k, n) = (11, 1), (11, 3), (19, 5), (25, 1), (25, 9), (49, 9), (53, 1), (97, 5), (107, 5), (131, 5), (137, 1), (149, 5), (257, 5), (277, 1), (313, 3)$, and $(421, 3)$. Solutions with $q + r = p^3$ and $p$ prime are $\sigma(2) + \sigma(6) = \sigma(8)$ and

$$\sigma(11638687) + \sigma(2^2 \times 13 \times 1123) = \sigma(227^3).$$

Erdös asks how many solutions (not necessarily primitive) are there with $q + r < x$; is it $cx + o(x)$ or is it of higher order? If $s_1 < s_2 < \cdots$ are the numbers for which $\sigma(s_i) = \sigma(q) + \sigma(s_i - q)$ has a solution with $q < s_i$, what is the density of the sequence $\{s_i\}$?

M. Sugunamma, PhD thesis, Sri Venkataswara Univ. 1969.

**B16.** Erdös and Szekeres studied numbers $n$ such that if a prime $p$ divides $n$, then $p^i$ divides $n$ where $i$ is a given number greater than one. Golomb named these numbers **powerful** and exhibited infinitely many pairs of consecutive ones. He conjectures that 6 is not representable as the difference of two powerful numbers and that there are infinitely many such non-representable numbers. On the other hand, Mąkowski has shown that every prime $\equiv 1 \pmod 8$ is the difference of two relatively prime powerful numbers, and Sentance has found an infinite number

$$(5^2 53^2, 3^5 17^2), \qquad (5^4 19^2 29^2, 3^3 11^2 241^2), \ldots$$

of pairs of consecutive odd powerful numbers, in addition to $(5^2, 3^3)$.

More generally, Erdös denotes by $u_1^{(k)} < u_2^{(k)} < \cdots$ the integers all of whose prime factors have exponents $\geq k$. He asks if the equation $u_{i+1}^{(2)} - u_i^{(2)} = 1$ has infinitely many solutions which do not come from Pell equations $x^2 - dy^2 = \pm 1$. Is there a constant $c$, such that the number of solutions with $u_i < x$ is less than $(\ln x)^c$? Does $u_{i+1}^{(3)} - u_i^{(3)} = 1$ have no solutions? Do the equations $u_{i+2}^{(2)} - u_{i+1}^{(2)} = 1 = u_{i+1}^{(2)} - u_i^{(2)}$ have no simultaneous solutions? What is the largest $r$ such that there are $u_{i_1}^{(2)}, \ldots, u_{i_r}^{(2)}$ in arithmetic progression? Are there infinitely many such $r$-tuples? Erdös conjectures that there are infinitely many triples of $u_i^{(3)}$ in arithmetic progression, but no quadruples, and no triples of $u_i^{(4)}$. Also that $u_i^{(3)} + u_j^{(3)} = u_k^{(3)}$ has infinitely many solutions, but that $u_i^{(4)} + u_j^{(4)} = u_k^{(4)}$ has at most a finite number. More generally that the sum of $k - 2$ of the $u_i^{(k)}$ is at most finitely often a $u_i^{(k)}$. Also that every sufficiently large integer is the sum of three $u_i^{(2)}$: he asks for a list of those integers which are not.

P. Erdös, Problems and results on consecutive integers, *Eureka* **38** (1975–76) 3–8.

P. Erdös and G. Szekeres, Über die Anzahl der Abelschen Gruppen gegebener Ordnung und über ein verwandtes zahlentheoretisches Problem, *Acta Litt. Sci. Szeged* **7** (1934) 95–102; *Zbl.* **10**, 294.

S. W. Golomb, Powerful numbers, *Amer. Math. Monthly* **77** (1970) 848–852; *MR* **42** #1780.

Andrzej Mąkowski, On a problem of Golomb on powerful numbers, *Amer. Math. Monthly* **79** (1972) 761.

W. A. Sentance, Occurences of consecutive odd powerful numbers, *Amer. Math. Monthly* **88** (1981) 272–274.

**B17.** If $n = p_1^{a_1} p_2^{a_2} \cdots p_r^{a_r}$, then Straus and Subbarao call $d$ an **exponential divisor** (e-divisor) of $n$ if $d \mid n$ and $d = p_1^{b_1} p_2^{b_2} \cdots p_r^{b_r}$ where $b_j \mid a_j$ $(1 \leq j \leq r)$, and they call $n$ **e-perfect** if $\sigma_e(n) = 2n$, where $\sigma_e(n)$ is the sum of the exponential divisors of $n$. Some examples of e-perfect numbers are

$$2^2 \times 3^2, \quad 2^2 \times 3^3 \times 5^2, \quad 2^3 \times 3^2 \times 5^2, \quad 2^4 \times 3^2 \times 11^2,$$
$$2^4 \times 3^3 \times 5^2 \times 11^2, \quad 2^6 \times 3^2 \times 7^2 \times 13^2, \quad 2^6 \times 3^3 \times 5^2 \times 7^2 \times 13^2,$$
$$2^7 \times 3^2 \times 5^2 \times 7^2 \times 13^2, \quad 2^8 \times 3^2 \times 5^2 \times 7^2 \times 139^2$$

and

$$2^{19} \times 3^2 \times 5^2 \times 7^2 \times 11^2 \times 13^2 \times 19^2 \times 37^2 \times 79^2 \times 109^2 \times 157^2 \times 313^2.$$

If $m$ is squarefree, $\sigma_e(m) = m$, so if $n$ is e-perfect and $m$ is squarefree with $(m, n) = 1$, then $mn$ is e-perfect. So it suffices to consider only powerful (B16) e-perfect numbers.

Straus and Subbarao show that there are no odd e-perfect numbers, in fact no odd $n$ which satisfy $\sigma_e(n) = kn$ for any integer $k > 1$. They also show that for each $r$ the number of (powerful) e-perfect numbers with $r$ prime factors is finite, and that the same holds for e-**multiperfect numbers** ($k > 2$).

Is there an e-perfect number which is *not* divisible by 3?

Straus and Subbarao conjecture that there is only a finite number of e-perfect numbers *not* divisible by any given prime $p$.

Are there any e-multiperfect numbers?

E. G. Straus and M. V. Subbarao, On exponential divisors, *Duke Math. J.* **41** (1974) 465–471; *MR* **50** #2053.
M. V. Subbarao, On some arithmetic convolutions, in *The Theory of Arithmetic Functions*, Springer-Verlag, New York, 1972.
M. V. Subbarao and D. Suryanarayana, Exponentially perfect and unitary perfect numbers, *Notices Amer. Math. Soc.* **18** (1971) 798.

**B18.** Is there an infinity of numbers $n$ such that $d(n) = d(n + 1)$? For example, $n = 2, 14, 21, 26, 33, 34, 38, 44, 57, 75, 85, 86, 93, 94, 98, 104, 116, 118$, 122, 133, 135, 141, 142, 145, 147, .... Many of these examples arise from pairs of consecutive numbers which are products of just two distinct primes, and it has even been conjectured that there is an infinity of *triples* of consecutive products of two primes, $n$, $n + 1$, $n + 2$. For example, $n = 33$, 85, 93, 141, 201, 213, 217, 301, 393, 445, 633, 697, 921, .... It is clearly not possible to have *four* such numbers, but it *is* possible to have longer sequences of consecutive numbers with the same number of divisors. For example,

$$d(242) = d(243) = d(244) = d(245) = 6$$

and

$$d(40311) = d(40312) = d(40313) = d(40314) = d(40315) = 8.$$

How long can such sequences be?

P. Erdös and L. Mirsky, The distribution of values of the divisor function $d(n)$, *Proc. London Math. Soc.* (3) **2** (1952) 257–271.
M. Nair and P. Shiu, On some results of Erdös and Mirsky, *J. London Math. Soc.* (2) **22** (1980) 197–203; and see *ibid.* **17** (1978) 228–230.
A. Schinzel, Sur un problème concernant le nombre de diviseurs d'un nombre naturel, *Bull. Acad. Polon. Sci. Ser. sci. math. astr. phys.* **6** (1958) 165–167.
A. Schinzel and W. Sierpiński, Sur certaines hypothèses concernant les nombres premiers, *Acta Arith.* **4** (1958) 185–208.
W. Sierpiński, Sur une question concernant le nombre de diviseurs premiers d'un nombre naturel, *Colloq. Math.* **6** (1958) 209–210.

**B19.** Motzkin and Straus asked for all pairs of numbers $m$, $n$ such that $m$ and $n+1$ have the same set of distinct prime factors, and similarly for $n$ and $m+1$. It was thought that such pairs were necessarily of the form $m = 2^k + 1$, $n = m^2 - 1$ $(k = 0, 1, 2, \ldots)$ until J. H. Conway observed that if $m = 5 \times 7$, $n + 1 = 5^4 \times 7$, then $n = 2 \times 3^7$, $m + 1 = 2^2 \times 3^2$. Are there others?

Similarly, Erdös asks if there are numbers $m, n$ $(m < n)$ other than $m = 2^k - 2$, $n = 2^k(2^k - 2)$ such that $m$ and $n$ have the same prime factors and similarly for $m+1$ and $n+1$. Mąkowski found the pair $m = 3 \times 5^2$, $n = 3^5 \times 5$ for which $m + 1 = 2^2 \times 19$, $n + 1 = 2^6 \times 19$.

Pomerance has asked if there are any odd numbers $n > 1$ such that $n$ and $\sigma(n)$ have the same prime factors. He conjectures that there are not.

A. Mąkowski, On a problem of Erdös, *Enseignement Math.* (2) **14** (1968) 193.

**B20.** Some interest has been shown in the **Cullen numbers**, $n \times 2^n + 1$, which are all composite for $2 \le n \le 1000$, except for $n = 141$. This is probably a good example of the strong law of small numbers, because for small $n$, where the density of primes is large, the Cullen numbers are very likely to be composite because Fermat's theorem tells us that $(p-1)2^{p-1} + 1$ and $(p-2)2^{p-2} + 1$ are both divisible by $p$.

Riesel observes that the corresponding numbers $n \times 2^n - 1$ are prime for $n = 2, 3, 6, 30, 75$, and 81, but for no others $\le 110$.

**B21.** Let $N(x)$ be the number of odd positive integers $k$, not exceeding $x$, such that $k \cdot 2^n + 1$ is prime for some positive integer $n$. Sierpiński used covering congruences (see F13) to show that $N(x)$ tends to infinity with $x$. For example, if

$$k \equiv 1 \pmod{641 \times (2^{32} - 1)} \quad \text{and} \quad k \equiv -1 \pmod{6700417},$$

then every member of the sequence $k \cdot 2^n + 1$ $(n = 0, 1, 2, \ldots)$ is divisible by at least one of the primes 3, 5, 17, 257, 641, 65537, and 6700417. He also noted that one of 3, 5, 7, 13, 17, 241 will always divide $k \cdot 2^n + 1$ for certain other values of $k$.

Erdös and Odlyzko have shown that

$$(\tfrac{1}{2} - c_1)x \ge N(x) \ge c_2 x.$$

What is the least value of $k$ such that $k \cdot 2^n + 1$ is composite for all values of $n$? Selfridge discovered that one of 3, 5, 7, 13, 19, 37, 73 always divides $78557 \times 2^n + 1$. He also noted that there is a prime of the form $k \cdot 2^n + 1$ for each $k < 383$, and that $383 \times 2^n + 1$ is composite for all $n < 2313$. N. S. Mendelsohn and B. Wolk extended this to $n \le 4017$, but more recently Hugh Williams discovered the prime $383 \times 2^{6393} + 1$.

It would seem that the determination of the least $k$ may now be within computer reach. Extensive calculations have been made by Baillie, Cormack, and Williams. After the discovery of several primes of the form $k \cdot 2^n + 1$, including those for which

$$k = 2897, 6313, 7493, 7957, 8543, 9323$$

and

$$n = 9715, 4606, 5249, 5064, 5793, 3013,$$

they were left with 118 candidates less than 78557. The first eight of these are

$$k = 3061, 4847, 5297, 5359, 5897, 7013, 7651, \text{and } 8423$$

for which no prime exists with

$$n \leq 16000, 8102, 8070, 8109, 8170, 8105, 8080, \text{and } 8000$$

respectively.

Robert Baillie, G. V. Cormack, and H. C. Williams, Some results concerning a problem of Sierpiński, *Math. Comput.*, (to appear).
G. V. Cormack and H. C. Williams, Some very large primes of the form $k \cdot 2^n + 1$, *Math. Comput.* **35** (1980) 1419–1421; *MR* **81i**: 10011.
P. Erdös and A. M. Odlyzko, On the density of odd integers of the form $(p-1)2^{-n}$ and related questions, *J. Number Theory* **11** (1979) 257–263.
J. L. Selfridge, Solution to problem 4995, *Amer. Math. Monthly* **70** (1963) 101.
W. Sierpiński, Sur un problème concernant les nombres $k \cdot 2^n + 1$, *Elem. Math.* **15** (1960) 73–74; *MR* **22** #7983; corrigendum, *ibid.* **17** (1962) 85.
W. Sierpiński, *250 Problems in Elementary Number Theory*, Elsevier, New York, 1970, 10, 64.

**B22.** Straus, Erdös, and Selfridge have asked that $n!$ be expressed as the product of $n$ factors, with the least one, $l$, as large as possible. For example, for $n = 56$, $l = 15$.

$$56! = 15 \times 16^3 \times 17^3 \times 18^8 \times 19^2 \times 20^{12} \times 21^9 \times 22^5 \times 23^2 \times 26^4 \times 29 \times 31 \times 37 \times 41 \times 43 \times 47 \times 53$$

Selfridge has two conjectures (a) that, except for $n = 56$, $l \geq \lfloor 2n/7 \rfloor$; (b) that for $n \geq 300000$, $l \geq n/3$ (if this is true, by how much can 300000 be reduced?) and Erdös, Selfridge and Straus have shown that for $n > n_0 = n_0(\varepsilon)$, $l > n/(e + \varepsilon)$. It is clear from Stirling's formula that this is best possible. Straus has shown that, by changing the positions of powers of 2 only, $l \geq 3n/16$. It is clear that $l$ is a monotonic increasing function of $n$ (though not, of course, strictly so). It does not, on the other hand, take all integer values: for $n = 124$, 125, $l$ is respectively 35 and 37. Erdös asks how large can the gaps be in the values of $l$, and can $l$ be constant for arbitrarily long stretches?

Alladi and Grinstead write $n!$ as a product of prime powers, each as large as $n^{\delta(n)}$ and let $\alpha(n) = \max \delta(n)$ and show that $\lim_{n\to\infty} \alpha(n) = e^{c-1} = \alpha$, say, where

$$c = \sum_{2}^{\infty} \frac{1}{k} \ln \frac{k}{k-1} \quad \text{so that} \quad \alpha = 0.809394020534 \ldots .$$

K. Alladi and C. Grinstead, On the decomposition of $n!$ into prime powers, *J. Number Theory* **9** (1977) 452–458.

P. Erdös, Some problems in number theory, *Computers in Number Theory*, Academic Press, London and New York, 1971, 405–414.

**B23.** Suppose $n! = a_1!a_2! \cdots a_r!$, $r \geq 2$, $a_1 \geq a_2 \geq \cdots \geq a_r \geq 2$. A trivial example is $a_1 = a_2! \cdots a_r! - 1$, $n = a_2! \cdots a_r!$ Dean Hickerson notes that the only nontrivial examples with $n \leq 410$ are $9! = 7!3!3!2!$, $10! = 7!6! = 7!5!3!$, and $16! = 14!5!2!$ and asks if there are any others.

Erdös observes that if $P(n)$ is the largest prime factor of $n$ and if it were known that $P(n(n+1))/\ln n$ tends to infinity with $n$, then it would follow that there are only a finite number of nontrivial examples.

He and Graham have studied the equation $y^2 = a_1!a_2! \cdots a_r!$. They define the set $F_k$ to be those $m$ for which there is a set of integers $m = a_1 > a_2 > \cdots > a_r$ with $r \leq k$ which satisfies this equation for some $y$, and write $D_k$ for $F_k - F_{k-1}$. They have various results, for example: for almost all primes $p$, $13p$ does not belong to $F_5$; and the least element of $D_6$ is 527. If $D_4(n)$ is the number of elements of $D_4$ which are $\leq n$, they do not know the order of growth of $D_4(n)$. They conjecture that $D_6(n) > cn$ but cannot prove this.

E. Ecklund and R. Eggleton, Prime factors of consecutive integers, *Amer. Math. Monthly* **79** (1972) 1082–1089.

E. Ecklund, R. Eggleton, P. Erdös and J. L. Selfridge, On the prime factorization of binomial coefficients, *J. Austral Math. Soc.* Ser. A, **26** (1978) 257–269; *MR* **80e**:10009.

P. Erdös, Problems and results on number theoretic properties of consecutive integers and related questions. *Congressus Numerantium XVI* Proc. 5th Manitoba Conf. Numer. Math. 1975, 25–44.

P. Erdös and R. L. Graham, On products of factorials, *Bull. Inst. Math. Acad. Sinica, Taiwan*, **4** (1976) 337–355.

**B24.** Let $f(n)$ be the size of the largest subset of $[1, n]$ no member of which divides two others. Erdös asks how large can $f(n)$ be? By taking $[m+1, 3m+2]$ it is clear that one can have $\lceil 2n/3 \rceil$. D. J. Kleitman shows that $f(29) = 21$ by taking $[11, 30]$ and omitting 18, 24 and 30, which then allows the inclusion of 6, 8, 9 and 10. However this example does not seem to generalize. In fact Lebensold has shown that if $n$ is large, then

$$0.6725n \leq f(n) \leq 0.6736n.$$

Dually, one can ask for the largest number of numbers $\leq n$, with no number a multiple of any two others. Kleitman's example serves this purpose also. More generally, Erdös asks for the largest number of numbers with no one divisible by $k$ others, for $k > 2$. For $k = 1$, the answer is $\lceil n/2 \rceil$.

Kenneth Lebensold, A divisibility problem, *Studies in Applied Math.* **56** (1976–77) 291–294. *MR* **58** #21639.

**B25.** Bateman asks if $31 = (2^5 - 1)/(2 - 1) = (5^3 - 1)/(5 - 1)$ is the only prime which is expressible in more than one way in the form $(p^r - 1)/(p^d - 1)$ where $p$ is a prime and $r \geq 3$ and $d \geq 1$ are integers. Trivially one has $7 = (2^3 - 1)/(2 - 1) = ((-3)^3 - 1)/(-3 - 1)$, but there are no others $< 10^{10}$. If the condition that $p$ be prime is relaxed, the problem goes back to Goormaghtigh and we have the solution $8191 = (2^{13} - 1)/(2 - 1) = (90^3 - 1)/(90 - 1)$.

E. T. Parker observes that the very long proof by Feit and Thompson, that every group of odd order is solvable, would be shortened if it could be proved that $(p^q - 1)/(p - 1)$ never divides $(q^p - 1)/(q - 1)$ where $p$, $q$ are distinct odd primes. In fact it has been conjectured that these two expressions are relatively prime, but Nelson M. Stephens has observed that when $p = 17$, $q = 3313$, they have a common factor $2pq + 1 = 112643$. McKay has established that $p^2 + p + 1 \nmid 3^p - 1$ for $p < 53 \times 10^6$.

P. T. Bateman and R. M. Stemmler, Waring's problem for algebraic number fields and primes of the form $(p^r - 1)/(p^d - 1)$, *Illinois J. Math.* **6** (1962) 142–156.
Ted Chinburg and Melvin Henriksen, Sums of $k$th powers in the ring of polynomials with integer coefficients, *Bull. Amer. Math. Soc.* **81** (1975) 107–110.
A. Mąkowski and A. Schinzel, Sur l'équation indéterminée de R. Goormaghtigh, *Mathesis* **68** (1959) 128–142 and **70** (1965) 94–96.
N. M. Stephens, On the Feit-Thompson conjecture, *Math. Comput.* **25** (1971) 625.

**B26.** Erdös asks what is the maximum $k$ so that the integers $a_i$, $1 \leq a_1 < a_2 < \cdots < a_k \leq n$ have no $l$ among them which are pairwise relatively prime. He conjectures that this is the number of integers $\leq n$ which have one of the first $l - 1$ primes as a divisor. He says that this is easy to prove for $l = 2$ and not difficult for $l = 3$, but he offers \$10.00 for a general solution.

Dually one can ask for the largest subset of $[1, n]$ whose members have pairwise l.c.m.'s not exceeding $n$. If $g(n)$ is the cardinality of such a maximal subset, then Erdös showed that

$$\frac{3}{2\sqrt{2}} n^{1/2} - 2 < g(n) \leq 2n^{1/2}$$

where the first inequality follows by taking the integers from 1 to $(n/2)^{1/2}$ together with the even integers from $(n/2)^{1/2}$ to $(2n)^{1/2}$. Choi improved the upper bound to $1.638n^{1/2}$.

S. L. G. Choi, The largest subset in $[1, n]$ whose integers have pairwise l.c.m. not exceeding $n$, *Mathematika* **19** (1972) 221–230.

S. L. G. Choi, On sequences containing at most three pairwise coprime integers, *Trans. Amer. Math. Soc.* **183** (1973) 437–440; *MR* **48** #6052.

P. Erdös, Extremal problems in number theory, *Proc. Sympos. Pure Math. Amer. Math. Soc.* **8** (1965) 181–189; *MR* **30** #4740.

**B27.** Erdös and Selfridge define $v(n;k)$ as the number of prime factors of $n+k$ which do not divide $n+i$ for $0 \le i < k$, and $v_0(n)$ as the maximum of $v(n;k)$ taken over all $k \ge 0$. Does $v_0(n) \to \infty$ with $n$? They show that $v_0(n) > 1$ for all $n$ except 1, 2, 3, 4, 7, 8, and 16. More generally, define $v_l(n)$ as the maximum of $v(n;k)$ taken over $k \ge l$. Does $v_l(n) \to \infty$ with $n$? They are unable to prove even that $v_1(n) = 1$ has only a finite number of solutions. Probably the greatest $n$ for which $v_1(n) = 1$ is $n = 330$.

They also denote by $V(n;k)$ the number of primes $p$ for which $p^\alpha$ is the highest power of $p$ dividing $n+k$, but $p^\alpha$ does not divide $n+i$ for $0 \le i < k$, and by $V_l(n)$ the maximum of $V(n;k)$ taken over $k \ge l$. Does $V_1(n) = 1$ have only a finite number of solutions? Perhaps $n = 80$ is the largest solution. What is the largest $n$ such that $V_0(n) = 2$?

Some further problems are given in their paper.

P. Erdös and J. L. Selfridge, Some problems on the prime factors of consecutive integers, *Illinois J. Math.* **11** (1967) 428–430.

A. Schinzel, Unsolved problem 31, *Elem. Math.* **14** (1959) 82–83.

**B28.** Selfridge asked: do there exist $n$ consecutive integers, each having either two distinct prime factors less than $n$ or a repeated prime factor less than $n$? He gives two examples:

(a) the numbers $a + 11 + i (1 \le i \le n = 115)$ where $a \equiv 0 \pmod{2^2 \times 3^2 \times 5^2 \times 7^2 \times 11^2}$ and $a + p \equiv 0 \pmod{p^2}$ for each prime $p$, $13 \le p \le 113$;
(b) the numbers $a + 31 + i\,(1 \le i \le n = 1329)$ where $a + p \equiv 0 \pmod{p^2}$ for each prime $p$, $37 \le p \le 1327$ and $a \equiv 0 \pmod{2^2 \times 3^2 \times 5^2 \times 7^2 \times 11^2 \times 13^2 \times 17^2 \times 19^2 \times 23^2 \times 29^2 \times 31^2}$.

It is harder to find examples of $n$ consecutive numbers, each one divisible by two distinct primes less than $n$ or by the square of a prime $< n/2$, though he believes they could be found, using a computer.

This is related to the problem: find $n$ consecutive integers, each having a composite common factor with the product of the other $n - 1$. If the composite condition is relaxed, and one asks merely for a common factor greater than 1, then $2184 + i\,(1 \le i \le n = 17)$ is a famous example.

Alfred Brauer, On a property of $k$ consecutive integers, *Bull. Amer. Math. Soc.* **47** (1941) 328–331; *MR* **2**, 248.

Ronald J. Evans, On blocks of $N$ consecutive integers, *Amer. Math. Monthly* **76** (1969) 48–49.

Ronald Evans, On $N$ consecutive integers in an arithmetic progression, *Acta Sci. Math. Univ. Szeged* **33** (1972) 295–296.

Heiko Harborth, Eine Eigenschaft aufeinanderfolgender Zahlen, *Arch. Math.* (Basel) **21** (1970) 50–51.
Heiko Harborth, Sequenzen ganzer Zahlen, in *Zahlentheorie, Berichte aus dem Math. Forschungsinst. Oberwolfach* **5** (1971) 59–66.
S. S. Pillai, On $m$ consecutive integers I, *Proc. Indian Acad. Sci.* Sect A, **11** (1940) 6–12; *MR* **1**, 199; II *ibid.* **11** (1940) 73–80; *MR* **1**, 291; III *ibid.* **13** (1941) 530–533; *MR* **3**, 66; IV *Bull. Calcutta Math. Soc.* **36** (1944) 99–101; *MR* **6**, 170.

**B29.** Alan R. Woods asks if there is a positive integer $k$ such that every $x$ is uniquely determined by the prime divisors of $x+1, x+2, \ldots, x+k$.

For primes less than 23 there are three ambiguous cases for $k = 2$: $(x, x+1) = (2,3)$ or $(8,9)$; $(14,15)$ or $(224,225)$; $(75,76)$ or $(1215,1216)$.

D. H. Lehmer, On a problem of Størmer, *Illinois J. Math.* **8** (1964) 57–79; *MR* **28** #2072.

**B30.** Erdös, Graham, and Selfridge want us to find the least value of $t_n$ so that the integers $n, n+1, \ldots, n+t_n$, contain a subset (of cardinality at least two) whose product is a square. The Thue–Siegel theorem implies that $t_n \to \infty$ with $n$, faster than a power of ln $n$.

Alternatively, is it true that for every $c$ there is an $n_0$ so that for every $n > n_0$ the products $\prod a_i$, taken over $n < a_1 < \cdots < a_k < n + (\ln n)^c$ ($k = 1, 2, \ldots$) are all distinct? They proved this for $c < 2$.

**B31.** Earl Ecklund, Roger Eggleton, Erdös, and Selfridge (see B23) write the **binomial coefficient** $\binom{n}{k} = n!/k!(n-k)!$ as a product $UV$ in which every prime factor of $U$ is at most $k$ and every prime factor of $V$ is greater than $k$. There are only finitely many cases with $n \geq 2k$ for which $U > V$. Determine all such cases when $k = 3$, 5 or 7.

S. P. Khare lists all cases with $n \leq 551$: $k = 3$, $n = 8, 9, 10, 18, 82$ and 162; $k = 5$, $n = 10$, 12 and 28; $k = 7$, $n = 21$, 30 and 54.

Erdös also observes that it is not known if $\binom{2n}{n}$ is ever squarefree for $n > 4$. Denote by $e = e(n)$ the largest exponent such that, for some prime $p$, $p^e$ divides $\binom{2n}{n}$. It is not known whether $e \to \infty$ with $n$. On the other hand he cannot disprove $e > c \ln n$.

Wolstenholme's theorem states that if $n$ is a prime $>3$, then

$$\binom{2n-1}{n} \equiv 1 \pmod{n^3}$$

James P. Jones asks if the converse is true.

P. Erdös and G. Szekeres, Some number theoretic problems on binomial coefficients, *Austral. Math. Soc. Gaz.* **5** (1978) 97–99; *MR* **80e**:10010.

**B32.** Grimm has conjectured that if $n+1, n+2, \ldots, n+k$ are all composite, then there are distinct primes $p_{i_j}$ such that $p_{i_j} | (n+j)$ for $1 \leq j \leq k$.

For example

$$1802\ 1803\ 1804\ 1805\ 1806\ 1807\ 1808\ 1809, \text{ and } 1810$$

are respectively divisible by

$$53\quad 601\quad 41\quad 19\quad 43\quad 139\quad 113\quad 67, \text{ and } 181$$

and

$$114\ 115\ 116\ 117\ 118\ 119\ 120\ 121\ 122\ 123\ 124\ 125, \text{ and } 126$$

by

$$19\quad 23\quad 29\quad 13\quad 59\quad 17\quad 2\quad 11\quad 61\quad 41\quad 31\quad 5, \text{ and } 7.$$

Erdös and Selfridge asked for an estimate of $f(n)$, the least number such that for each $m$ there are distinct integers $a_1, a_2, \ldots, a_{\pi(n)}$ in $[m+1, m+f(n)]$ with $p_i | a_i$ where $p_i$ is the $i$th prime. They and Pomerance show that, for large $n$,

$$(3-\varepsilon)n \le f(n) \ll n^{3/2}(\ln n)^{-1/2}.$$


P. Erdös, Problems and results in combinatorial analysis and combinatorial number theory, *Congressus Numerantium* XXI in *Proc. 9th S.E. Conf. Combin. Graph Theory, Comput.*, Boca Raton, Utilitas Math. Winnipeg, 1978, 29–40.

P. Erdös and C. Pomerance, Matching the natural numbers up to $n$ with distinct multiples in another interval, *Nederl. Akad. Wetensch. Proc. Ser.* A **83** (=*Indag. Math.* **42**) (1980) 147–161.

Paul Erdös and Carl Pomerance, An analogue of Grimm's problem of finding distinct prime factors of consecutive integers, *Utilitas Math.* **19** (1981), (to appear).

P. Erdös and J. L. Selfridge, Some problems on the prime factors of consecutive integers II, in *Proc. Washington State Univ. Conf. Number Theory*, Pullman, 1971, 13–21.

C. A. Grimm, A conjecture on consecutive composite numbers, *Amer. Math. Monthly* **76** (1969) 1126–1128.

Michel Langevin, Plus grand facteur premier d'entiers en progression arithmétique, Sém. Delange-Pisot-Poitou, **18** (1976/77) Théorie des nombres, Fasc. 1, Exp. No. 3 (1977); *MR* **81a**:10011.

Carl Pomerance, Some number theoretic matching problems, in *Proc. Number Theory Conf.*, Queens Univ., Kingston, 1979, 237–247.

K. T. Ramachandra, N. Shorey, and R. Tijdeman, On Grimm's problem relating to factorization of a block of consecutive integers, *J. reine angew. Math.* **273** (1975) 109–124.


**B33.** What can one say about the largest divisor of the binomial coefficient $\binom{n}{k} = n!/k!(n-k)!$ which is less than $n$? Erdös points out that it is easy to show that it is at least $n/k$ and conjectures that there may be one between $cn$ and $n$ for any $c < 1$ and $n$ sufficiently large. Marilyn Faulkner showed that if $p$ is the least prime $>2k$ and $n \ge p$, then $\binom{n}{k}$ has a prime divisor $\ge p$, except for $\binom{9}{2}$ and $\binom{10}{3}$. Earl Ecklund showed that if $n \ge 2k > 2$ then $\binom{n}{k}$ has a prime divisor $p \le n/2$, except for $\binom{7}{3}$.

John Selfridge conjectures that if $n \ge k^2 - 1$, then, apart from the exception $\binom{62}{6}$, there is a prime divisor $\le n/k$ of $\binom{n}{k}$. In the other direction he conjectures that if $n < k(k+3)$, then there is a prime divisor $\le k+1$ of $\binom{n}{k}$, apart from the five exceptions $\binom{7}{3}$ $\binom{14}{4}$ $\binom{23}{5}$ $\binom{44}{8}$ and $\binom{47}{11}$. These exceptions would be covered if the conjecture were weakened to $\le k+3$.

A classical theorem discovered independently by Sylvester and Schur, states that the product of $k$ consecutive integers, each greater than $k$, has a prime divisor greater than $k$. Leo Moser conjectured that the Sylvester-Schur theorem holds for primes $\equiv 1 \pmod 4$ i.e., that such a product of $k$ consecutive integers has a prime divisor $\equiv 1 \pmod 4$ which is greater than $k$. However Erdös does not think that this is true.

Neil Sloane observes that $(3/(5m+3))\binom{5m+3}{m}$ is always an integer and asks for a generalization to $(a/r)\binom{n}{r}$. Generalizations may exist of which the prototype is the **Catalan numbers** $(1/(n+1))\binom{2n}{n}$.

$f(n)$ is the sum of the reciprocals of those primes $< n$ which do not divide $\binom{2n}{n}$. Erdös et al conjectured that there is an absolute constant $c$ so that $f(n) < c$ for all $n$. Erdös also conjectured that for $n > 4$, $\binom{2n}{n}$ is never squarefree. Since $4 \mid \binom{2n}{n}$ unless $n = 2^k$, it suffices to consider

$$\binom{2^{k+1}}{2^k}.$$

Erdös has also conjectured that for $k > 8$, $2^k$ is not the sum of distinct powers of 3 $[2^8 = 3^5 + 3^2 + 3 + 1]$. If that's true, then for $k \ge 9$,

$$3 \,\Big|\, \binom{2^{k+1}}{2^k}.$$

Is $\binom{342}{171}$ the largest $\binom{2n}{n}$ which is not divisible by the square of an odd prime?

Graham offers \$100.00 for deciding if $(\binom{2n}{n}, 105) = 1$ infinitely often. Kummer knew that $n$, when written in base 3, 5, or 7, would have to have only the digits 0, 1; 0, 1, 2; or 0, 1, 2, 3 respectively. H. Gupta and S. P. Khare found the 14 values of $n$ less then $7^{10}$: 1, 10, 756, 757, 3160, 3186, 3187, 3250, 7560, 7561, 7651, 20007, 59548377, and 59548401. Peter Montgomery, Khare, and others found many larger values.

Graham, Erdös, Ruzsa, and Straus showed that for any *two* primes $p$, $q$ there are infinitely many $n$ for which $(\binom{2n}{n}, pq) = 1$.

If $g(n)$ is the smallest prime factor of $\binom{2n}{n}$, then $g(3160) = 13$ and $g(n) \le 11$ for $3160 < n < 10^{110}$.

E. F. Ecklund, On prime divisors of the binomial coefficient, *Pacific J. Math.* **29** (1969) 267–270.

P. Erdös, A theorem of Sylvester and Schur, *J. London Math. Soc.* **9** (1934) 282–288.

Paul Erdös, A mélange of simply posed conjectures with frustratingly elusive solutions, *Math. Mag.* **52** (1979) 67–70.

P. Erdös and R. L. Graham, On the prime factors of $\binom{n}{k}$, *Fibonacci Quart.* **14** (1976) 348–352.

P. Erdös, R. L. Graham, I. Z. Ruzsa, and E. Straus, On the prime factors of $\binom{2n}{n}$, Math. Comput. **29** (1975) 83–92.

M. Faulkner, On a theorem of Sylvester and Schur, *J. London Math. Soc.* **41** (1966) 107–110.

L. Moser, Insolvability of $\binom{2n}{n} = \binom{2a}{a}\binom{2b}{b}$, *Canad. Math. Bull.* **6** (1963) 167–169.
I. Schur, Einige Sätze über Primzahlen mit Anwendungen and Irreduzibilitätsfragen I, *S.-B. Preuss. Akad. Wiss. Phys.-Math. Kl.* **14** (1929) 125–136.
J. Sylvester, On arithmetical series, *Messenger of Math.* **21** (1892) 1–19, 87–120.
W. Utz, A conjecture of Erdös concerning consecutive integers, *Amer. Math. Monthly* **68** (1961) 896–897.

**B34.** If $H_{k,n}$ is the proposition: there is an $i$, $0 \le i < k$ such that $n - i$ divides $\binom{n}{k}$, then Erdös asked if $H_{k,n}$ is true for all $k$ when $n \ge 2k$. Schinzel gave the counter-example $n = 99215$, $k = 15$. If $H_k$ is the proposition: $H_{k,n}$ is true for all $n$, then Schinzel showed that $H_k$ is false for $k = 15$, 21, 22, 33, 35, 45, 55, 63, 65, 69, 75, 77, 85, 87, 91, 93, 95, and 99. He asked if there are infinitely many such $k$ and Erdös showed that there are. Schinzel showed that $H_k$ is true for all other $k \le 32$ and asked if there are infinitely many $k$, other than prime-powers, for which $H_k$ is true: he conjectures not.

A. Schinzel, Sur un problème de P. Erdös, *Colloq. Math.* **5** (1957–58) 198–204.

**B35.** Let $f(n)$ be the least integer such that at least one of the numbers $n, n+1, \ldots, n+f(n)$ divides the product of the others. It is easy to see that $f(k!) = k$, and $f(n) > k$ for $n > k!$. Erdös has also shown that

$$f(n) > \exp((\ln n)^{1/2-\varepsilon})$$

for an infinity of values of $n$, but it seems difficult to find a good upper bound for $f(n)$.

Erdös asks if $(m+1)(m+2)\cdots(m+k)$ and $(n+1)(n+2)\cdots(n+l)$ with $k \ge l \ge 3$ can contain the same prime factors infinitely often. For example $(2 \times 3 \times 4 \times 5 \times 6) \times 7 \times 8 \times 9 \times 10$ and $14 \times 15 \times 16$ and $48 \times 49 \times 50$; also $(2 \times 3 \times 4 \times 5 \times 6) \times 7 \times 8 \times 9 \times 10 \times 11 \times 12$ and $98 \times 99 \times 100$. For $k = l \ge 3$ he conjectures that this happens only finitely many times.

If $L(n;k)$ is the l.c.m. of $n+1, n+2, \ldots, n+k$, then Erdös conjectures that for $l > 1$, $n \ge m+k$, $L(m;k) = L(n;l)$ has only a finite number of solutions. Examples are $L(4;3) = L(13;2)$ and $L(3;4) = L(19;2)$. He asks if there are infinitely many $n$ such that for all $k$ $(1 \le k < n)$ we have $L(n;k) > L(n-k;k)$. What is the largest $k = k(n)$ for which this inequality can be reversed? He notes that it is easy to see that $k(n) = o(n)$, but believes that much more is true. He expects that for every $\varepsilon > 0$ and $n > n_0(\varepsilon)$, $k(n) < n^{1/2+\varepsilon}$, but cannot prove this.

P. Erdös, How many pairs of products of consecutive integers have the same prime factors? *Amer. Math. Monthly* **87** (1980) 391–392.

**B36.** Euler's **totient function**, $\phi(n)$, is the number of numbers not greater than $n$ and prime to $n$. For example $\phi(1) = \phi(2) = 1$, $\phi(3) = \phi(4) = \phi(6) = 2$, $\phi(5) = \phi(8) = \phi(10) = \phi(12) = 4$, $\phi(7) = \phi(9) = 6$. Are there infinitely many pairs of consecutive numbers, $n$, $n+1$, such that $\phi(n)=\phi(n+1)$? For example, $n = 1, 3, 15, 104, 164, 194, 255, 495, 584, 975$. It is not even known if $|\phi(n+1)-\phi(n)| < n^{\varepsilon}$ has an infinity of solutions for each $\varepsilon > 0$.

Schinzel conjectures that for every even $k$ the equation $\phi(n+k)=\phi(n)$ has an infinity of solutions. He observes that the corresponding conjecture with $k$ odd is implausible. For $k=1$, the problem of the previous paragraph, there are 18 solutions $<10^4$, while for $k=3$ there are only the solutions $n=3$ and $n=5$ in the same range. D. H. Lehmer has extended the range to $10^6$, revealing 59 solutions for $k=1$, but still only two for $k=3$. Sierpiński has shown that $\phi(n+k)=\phi(n)$ has at least one solution for each value of $k$ and Schinzel and Wakulicz have shown that there are at least two solutions for each $k < 2\times 10^{58}$. Mąkowski has shown that $\phi(n+k)=2\phi(n)$ has at least one solution for every $k$.

Two curiosities are $\phi(25930)=\phi(25935)=\phi(25940)=\phi(25942)=2^7 3^4$ and $\phi(404471)=\phi(404473)=\phi(404477)=2^8\times 3^2\times 5^2\times 7$.

**Nontotients** are positive values of $n$ for which $\phi(x)=n$ has no solution; for example $n = 14, 26, 34, 38, 50, 62, 68, 74, 76, 86, 90, 94, 98$. The number, $\#(y)$, of these less than $y$ have been calculated by the Lehmers.

| $y$ | $10^3$ | $10^4$ | $2\times 10^4$ | $3\times 10^4$ | $4\times 10^4$ | $5\times 10^4$ | $6\times 10^4$ | $7\times 10^4$ | $8\times 10^4$ | $9\times 10^4$ |
|---|---|---|---|---|---|---|---|---|---|---|
| $\#(y)$ | 210 | 2627 | 5515 | 8458 | 11438 | 14439 | 17486 | 20536 | 23606 | 26663 |

Erdös and Hall have shown that the number $\Phi(y) = y - \#(n)$, of $n$ for which $\phi(x)=n$ *has* a solution is $ye^{f(y)}/\ln y$, where $f(y)$ lies between $c(\ln\ln\ln y)^2$ and $c(\ln y)^{1/2}$. Erdös conjectures that $\Phi(cy)/\Phi(y)\to c$, and that this, if true, may be the best substitute that one can find for an asymptotic formula for $\Phi(y)$.

**Noncototients** are positive values of $n$ for which $x-\phi(x)=n$ has no solution; for example $n = 10, 26, 34, 50, 52, 58, 86, 100$. Sierpiński and Erdös conjecture that there are infinitely many noncototients.

P. Erdös, Über die Zahlen der Form $\sigma(n)-n$ and $n-\phi(n)$, *Elem. Math.* **28** (1973) 83–86.

P. Erdös and R. R. Hall, Distinct values of Euler's $\phi$-function, *Mathematika* **23** (1976) 1–3.

Andrzej Mąkowski, On the equation $\phi(n+k)=2\phi(n)$, *Elem. Math.* **29** (1974) 13.

A. Schinzel, Sur l'équation $\phi(x+k)=\phi(x)$, *Acta Arith.* **4** (1958) 181–184; *MR* **21** #5597.

A. Schinzel and A. Wakulicz, Sur l'équation $\phi(x+k)=\phi(x)$ II, *Acta Arith.* **5** (1959) 425–426; *MR* **23** #A831.

W. Sierpiński, Sur un propriété de la fonction $\phi(n)$, *Publ. Math. Debrecen* **4** (1956) 184–185.

**B37.** D. H. Lehmer has conjectured that there is no composite value of $n$ such that $\phi(n)$ is a divisor of $n-1$, i.e., that for no value of $n$ is $\phi(n)$ a *proper* divisor of $n-1$. Such an $n$ must be a Carmichael number (A13). He showed that it would have to be the product of at least seven distinct primes, and Lieuwens has proved the following theorems. If $3|n$, $n$ is the product of more than 212 primes and $n > 5 \cdot 5 \times 10^{571}$. If the smallest prime factor of $n$ is 5, $n$ contains at least 11 prime factors. If the smallest prime factor of $n$ is at least 7, then $n$ is the product of at least 13 primes. This supersedes and corrects the work of Schuh. Masao Kishore has shown that at least 13 primes are needed in any case and Cohen and Hagis have improved this to 14. Pomerance has proved that the number of composite $n$ less than $x$ for which $\phi(n)|(n-1)$ is less than $x^{1/2+\varepsilon}$.

Schinzel notes that if $n = p$ or $2p$, where $p$ is prime, then $\phi(n)+1$ divides $n$, and asks if the converse is always true.

If $n$ is prime, it divides $\phi(n)d(n)+2$. Is this true for any composite $n$, other than $n = 4$? Subbarao also notes that if $n$ is prime, then $n\sigma(n) \equiv 2$, mod $\phi(n)$, also if $n = 4, 6, 22$; is it true for infinitely many composite $n$?

Subbarao has an analogous conjecture to Lehmer's, based on the function $\phi^*(n) = \prod(p^a - 1)$, where the product is taken over the maximal prime powers that divide $n$, $p^a \| n$. He conjectures that if $\phi^*(n)|(n-1)$, then $n$ is a power of a prime.

Ron Graham makes the following conjecture

¿ For all $k$ there are infinitely many $n$ such that $\phi(n)|(n-k)$ ?

He observes that it is true for $k = 0$, $k = 2^a$ $(a \geq 0)$ and $k = 2^a 3^b$ $(a, b > 0)$ for example.


Ronald Alter, Can $\phi(n)$ properly divide $n-1$? *Amer. Math. Monthly* **80** (1973) 192–193.

Graeme L. Cohen and Peter Hagis, On the number of prime factors of $n$ if $\phi(n)|(n-1)$, *Nieuw Arch. Wisk.* (3) **28** (1980), 177–185.

Masao Kishore, On the equation $k\phi(M) = M - 1$, *Nieuw Arch. Wisk.* (3) **25** (1977) 48–53. See also *Notices Amer. Math. Soc.* **22** (1975) A-501–502.

D. H. Lehmer, On Euler's totient function, *Bull. Amer. Math. Soc.* **38** (1932) 745–751.

E. Lieuwens, Do there exist composite numbers for which $k\phi(M) = M - 1$ holds? *Nieuw Arch. Wisk.* (3) **18** (1970) 165–169; *MR* **42** #1750.

C. Pomerance, On composite $n$ for which $\phi(n)|n-1$, *Acta Arith.* **28** (1976) 387–389; II, *Pacific J. Math.* **69** (1977) 177–186. See also *Notices Amer. Math. Soc.* **22** (1975) A-542.

Fred Schuh, Can $n-1$ be divisible by $\phi(n)$ where $n$ is composite? *Mathematica*, Zutphen B. **12** (1944) 102–107.

M. V. Subbarao, On two congruences for primality, *Pacific J. Math.* **52** (1974) 261–268; *MR* **50** #2049.

David W. Wall, Conditions for $\phi(N)$ to properly divide $N-1$, *A Collection of Manuscripts Related to the Fibonacci Sequence*, 18th Anniv. Vol., Fibonacci Assoc. 205–208.


**B38.** Are there infinitely many pairs of numbers $m, n$ such that $\phi(m) = \sigma(n)$? Since, for $p$ prime, $\phi(p) = p - 1$ and $\sigma(p) = p + 1$, this question would be answered affirmatively if there were infinitely many twin primes (A7). Also if there were infinitely many Mersenne primes (A3) $M_p = 2^p - 1$, since

$\sigma(M_p) = 2^p = \phi(2^{p+1})$. However, there are many solutions other than these, sometimes displaying little noticeable pattern, e.g. $\phi(780) = 192 = \sigma(105)$.

Erdös remarks that the equation $\phi(x) = n!$ is solvable, and (apart from $n = 2$) $\sigma(y) = n!$ is probably solvable also.

**B39. Carmichael's conjecture.** For every $n$ it appears to be possible to find an $m$, not equal to $n$, such that $\phi(m) = \phi(n)$ and for a few years early in this century it was thought that Carmichael had proved this. Klee verified the conjecture for $\phi(n) < 10^{400}$, and for all $n$ not divisible by $2^{42} \times 3^{47}$. Pomerance has shown that if $n$ is such that for every prime $p$ for which $p - 1$ divides $\phi(n)$ we have $p^2$ divides $n$, then $n$ is a counter-example. He can also show (unpublished) that a conjecture of Schinzel, that the first prime $p \equiv 1 \pmod q$ is less than $q^2$, implies that there are no numbers $n$ which satisfy his theorem.

Erdös proved that if $\phi(x) = k$ has exactly $s$ solutions, then there are infinitely many $k$ for which there are exactly $s$ solutions, and that $s > k^c$ for infinitely many $k$. If $C$ is the least upper bound of those $c$ for which this is true, then Wooldridge showed that $C \geq 3 - 2\sqrt{2} > 0.17157$. Pomerance used Hooley's improvement on the Brun–Titchmarsh theorem to improve this to $C \geq 1 - 625/512e > 0.55092$ and notes that recent further improvements by Iwaniec enable him to get $C > 0.55655$ so that $s > k^{5/9}$ for infinitely many $k$. Erdös conjectures that $C = 1$. In the other direction Pomerance also shows that

$$s < k \exp\{-(1 + o(1))\ln k \ln\ln\ln k/\ln\ln k\}$$

and gives a heuristic argument to support the belief that this is best possible.

R. D. Carmichael, Note on Euler's $\phi$-function, *Bull. Amer. Math. Soc.* **28** (1922) 109–110.

P. Erdös, on the normal number of prime factors of $p - 1$ and some other related problems concerning Euler's $\phi$-function, *Quart. J. Math. Oxford Ser.* **6** (1935) 205–213.

P. Erdös, Some remarks on Euler's $\phi$-function and some related problems, *Bull. Amer. Math. Soc.* **51** (1945) 540–544.

P. Erdös, Some remarks on Euler's $\phi$ function, *Acta Arith.* **4** (1958) 10–19; *MR* **22** #1539.

P. Erdös and R. R. Hall. Distinct values of Euler's $\phi$-function, *Mathematika* **23** (1976) 1–3.

C. Hooley, On the greatest prime factor of $p + a$, *Mathematika* **20** (1973) 135–143.

H. Iwaniec, On the Brun-Titchmarsh theorem, (to appear)

V. L. Klee, On a conjecture of Carmichael, *Bull. Amer. Math. Soc.* **53** (1947) 1183–1186; *MR* **9**, 269.

V. L. Klee, Is there an $n$ for which $\phi(x) = n$ has a unique solution? *Amer. Math. Monthly* **76** (1969) 288–289.

Carl Pomerance, On Carmichael's conjecture, *Proc. Amer. Math. Soc.* **43** (1974) 297–298.

Carl Pomerance, Popular values of Euler's function, *Mathematika* **27** (1980) 84–89.

K. R. Wooldridge, Values taken many times by Euler's phi-function, *Proc. Amer. Math. Soc.* **76** (1979) 229–234; *MR* **80g**:10008.

**B40.** If $a_1 < a_2 < \cdots < a_{\varphi(n)}$ are the integers less than $n$ and prime to it, Erdös conjectures that $\sum(a_{i+1} - a_i)^2 < cn^2/\phi(n)$ and offers \$100.00 for a proof. Hooley has shown that, for $1 \le \alpha < 2$, $\sum(a_{i+1} - a_i)^\alpha \ll n(n/\phi(n))^{\alpha-1}$ and that $\sum(a_{i+1} - a_i)^2 \ll n(\ln\ln n)^2$, while Vaughan has established the conjecture "on the average" by showing that

$$\sum(a_{i+1} - a_i)^2 \ll n^2(1 + \sum_{p|n} (\ln p)/p)/\phi(n).$$

Jacobsthal asked what bounds can be placed on $\max(a_{i+1} - a_i)$.

P. Erdös, On the integers relatively prime to $n$ and on a number-theoretic function considered by Jacobsthal, *Math. Scand.* **10** (1962) 163–170; *MR* **26** #3651.

C. Hooley, On the difference of consecutive numbers prime to $n$, *Acta Arith.* **8** (1963) 343–347.

R. C. Vaughan, Some applications of Montgomery's sieve, *J. Number Theory* **5** (1973) 64–79.

**B41.** There is a close relative to the sum of divisors and the sum of the unitary divisors function, which complements Euler's totient function. If $n = p_1^{a_1}p_2^{a_2} \cdots p_k^{a_k}$, denote by $\bar{\phi}(n)$ the product $\prod p_i^{a_i-1}(p_i + 1)$, i.e., $\bar{\phi}(n) = n\prod(1 + p^{-1})$, where the product is taken over the distinct prime divisors of $n$. It is easy to see that iteration of the function leads eventually to terms of the form $2^a \times 3^b$ where $b$ is fixed and $a$ increases by one in successive terms. Given any value of $b$ there are infinitely many initial values of $n$ which lead to such terms, for example, $\bar{\phi}^k(2^a \times 3^b \times 7^c) = 2^{a+4k} \times 3^b \times 7^{c-k}$ $(0 \le k \le c)$ and $\bar{\phi}^k(2^a \times 3^b \times 7^c) = 2^{a+5k-c} \times 3^b$ $(k > c)$.

David E. Penney and Pomerance, in an unpublished paper, show that there are values of $n$ for which the iterates of the function $\bar{\phi}(n) - n$ are unbounded as the number of iterations tends to infinity; the least such is $n = 318$.

If we average $\bar{\phi}$ with the $\phi$-function, $\frac{1}{2}(\phi + \bar{\phi})$, and iterate, we produce sequences whose terms become constant whenever they are prime powers; for example 24, $\frac{1}{2}(8+48)=28$, $\frac{1}{2}(12+48)=30$, $\frac{1}{2}(8+72)=40$, $\frac{1}{2}(16+72)=44$, $\frac{1}{2}(20+72)=46$, $\frac{1}{2}(22+72)=47$, $\frac{1}{2}(46+48)=47, \ldots$ . Are there others which increase indefinitely?

We can also average the $\sigma$- and $\phi$-functions, and iterate. Since $\phi(n)$ is always even for $n > 2$ and $\sigma(n)$ is odd when $n$ is a square or twice a square, we will sometimes get a noninteger value. For example 54, 69, 70, 84, 124, 142, 143, 144, $225\frac{1}{2}$; in this case we say the sequence **fractures**. It is easy to show that $(\sigma(n) + \phi(n))/2 = n$ just if $n = 1$ or a prime, so sequences can become constant, for example, 60, 92, 106, 107, 107, $\ldots$ . Once again, are there sequences which increase indefinitely?

Of course, if we iterate the $\phi$-function, it eventually arrives at 2. Call the integer $k$ for which $\phi^k(n) = 2$ the **class** of $n$.

| $k$ | $n$ | |
|---|---|---|
| 0 | | 2 |
| 1 | 3 | 4 6 |
| 2 | 5 7 | 8 9 10 12 14 18 |
| 3 | 11 13 15 | 16 19 20 21 22 24 26 . . . |
| 4 | 17 23 25 29 31 | 32 33 34 35 37 39 40 43 . . . |
| 5 | 41 47 51 53 55 59 61 | 64 65 67 68 69 71 73 . . . |
| 6 | 83 85 89 97 101 103 107 113 115 119 121 122 123 125 128 . . . | |

$M = \{2, 3, 5, 11, 17, 41, 83, \ldots\}$ is the set of least values of the classes. Shapiro conjectured that $M$ contained only prime values, but Mills found several composite members. If $S$ is the union, for all $k$, of the members of class $k$ which are $< 2^{k+1}$,

$$S = \{3; 5, 7; 11, 13, 15; 17, 23, 25, 29, 31; 41, 47, 51, 53, 55, 59, 61; 83, 85, \ldots\},$$

then Shapiro showed that the factors of an element of $S$ are also in $S$. Catlin showed that if $m$ is an odd element of $M$, then the factors of $m$ are in $M$, and that there are finitely many primes in $M$ just if there are finitely many odd numbers in $M$. Does $S$ contain infinitely many odd numbers? Does $M$ contain infinitely many odd numbers?

Finucane iterated the function $\phi(n) + 1$ and asked: in how many steps does one reach a prime? Also, given a prime $p$, what is the distribution of the values of $n$ whose sequence ends with $p$. Are 5, 8, 10, 12 the only numbers which lead to 5? And 7, 9, 14, 15, 16, 18, 20, 24, 30 the only ones leading to 7?

Erdös similarly asked about the iteration of $\sigma(n) - 1$. Does it always end on a prime, or can it grow indefinitely? In none of the cases of iteration of $\sigma(n) - 1$, of $(\overline{\phi}(n) + \phi(n))/2$, or of $(\phi(n) + \sigma(n))/2$ is he able to show that the growth is slower than exponential.

P. A. Catlin, Concerning the iterated $\phi$-function, *Amer. Math. Monthly* **77** (1970) 60–61.
Paul Erdös and R. R. Hall, Euler's $\phi$-function and its iterates, *Mathematika* **24** (1977) 173–177; *MR* **57** #12356.
W. H. Mills, Iteration of the $\phi$-function, *Amer. Math. Monthly* **50** (1953) 547–549.
C. A. Nicol, Some diophantine equations involving arithmetic functions, *J. Math. Anal. Appl.* **15** (1966) 154–161.
Harold N. Shapiro, An arithmetic function arising from the $\phi$-function, *Amer. Math. Monthly* **50** (1943) 18–30; *MR* **4**, 188.

**B42.** Mąkowski and Schinzel prove that $\limsup \phi(\sigma(n))/n = \infty$,

$$\limsup \phi^2(n)/n = \frac{1}{2}, \quad \text{and} \quad \liminf \sigma(\phi(n))/n \le \frac{1}{2} + \frac{1}{2^{34} - 4}$$

and they ask if $\sigma(\phi(n))/n \ge \frac{1}{2}$ for all $n$. They point out that even $\inf \sigma(\phi(n))/n > 0$ is not proved.

A. Mąkowski and A. Schinzel, On the functions $\phi(n)$ and $\sigma(n)$, *Colloq. Math.* **13** (1964–65) 95–99.

**B43.** The numbers

$$\begin{aligned} &3! - 2! + 1! = 5,\\ &4! - 3! + 2! - 1! = 19,\\ &5! - 4! + 3! - 2! + 1! = 101,\\ &6! - 5! + 4! - 3! + 2! - 1! = 619,\\ &7! - 6! + 5! - 4! + 3! - 2! + 1! = 4421\\ \text{and}\quad &8! - 7! + 6! - 5! + 4! - 3! + 2! - 1! = 35899 \end{aligned}$$

are each prime. Are there infinitely many such? Here are the factors of $A_n = n! - (n-1)! + (n-2)! - + \cdots - (-1)^n 1!$ for the next few values of $n$.

| $n$ | $A_n$ | $n$ | $A_n$ |
|---|---|---|---|
| 9 | $79 \times 4139$ | 19 | 15578717622022981 (prime) |
| 10 | 3301819 (prime) | 20 | $8969 \times 210101 \times 1229743351$ |
| 11 | $13 \times 2816537$ | 21 | $113 \times 167 \times 4511191 \times 572926421$ |
| 12 | $29 \times 15254711$ | 22 | $79 \times 239 \times 56947572104043899$ |
| 13 | $47 \times 1427 \times 86249$ | 23 | $85439 \times 289993909455734779$ |
| 14 | $211 \times 1679 \times 229751$ | 24 | $12203 \times 24281 \times 2010359484638233$ |
| 15 | 1226280710981 (prime) | 25 | $59 \times 555307 \times 455254005662640637$ |
| 16 | $53 \times 6581 \times 56470483$ | 26 | $1657 \times 234384986539153832538067$ |
| 17 | $47 \times 7148742955723$ | 27 | $127^2 \times 271 \times 1163 \times 2065633479970130593$ |
| 18 | $2683 \times 2261044646593$ | 28 | $61 \times 221171 \times 21820357757749410439949$ |

The example $n = 27$ shows that these numbers are not necessarily squarefree.

If there is a value of $n$ such that $n + 1$ divides $A_n$, then $n + 1$ will divide $A_m$ for all $m > n$, and there would only be a finite number of prime values. Wagstaff established that if there is such an $n$, it is larger than 46340.

**B44.** Đ. Kurepa defines $!n = 0! + 1! + 2! + \cdots + (n-1)!$ and asks if $!n \not\equiv 0 \pmod n$ for all $n > 2$. Slavić used a computer to establish this for $3 \le n \le 1000$. The conjecture is that $(!n, n!) = 2$. Wagstaff has extended the calculations and verified the conjecture for $n < 50{,}000$. He notes that for $B_n = !(n+1) - 1 = 1! + 2! + \cdots + n!$ we have $3 \mid B_n$ for $n \ge 2$, $9 \mid B_n$ for $n \ge 5$ and $99 \mid B_n$ for $n \ge 10$.

L. Carlitz, A note on the left factorial function, *Math. Balkanica* **5** (1975) 37–42.
Đuro Kurepa, On some new left factorial propositions, *Math. Balkanica* **4** (1974) 383–386; *MR* **58** #10716.

**B45.** The coefficients in the expansion of $\sec x = \sum E_n x^n/n!$ are the **Euler numbers**, and arise in several combinatorial contexts. $E_0 = 1$, $E_2 = -1$, $E_4 = 5$, $E_6 = -61$, $E_8 = 1385$, $E_{10} = -50521$, $E_{12} = 2702765$, $E_{14} = -199360981$, $E_{16} = 19391512145$, $E_{18} = -2404879675441, \ldots$. Is it true that for any prime $p \equiv 1 \pmod 8$, $E_{(p-1)/2} \not\equiv 0 \pmod p$? It is true for $p \equiv 5 \pmod 8$.

E. Lehmer, On congruences involving Bernoulli numbers and the quotients of Fermat and Wilson, *Annals of Math.* **39** (1938) 350–360; *Zbl.* **19**, 5.
Barry J. Powell, Advanced problem 6325, *Amer. Math. Monthly* **87** (1980) 826.

**B46.** Erdös denotes by $P(n)$ the largest prime factor of $n$ and asks if there are infinitely many primes $p$ such that $(p-1)/P(p-1) = 2^k$? Or $= 2^k \times 3^l$?

Erdös and Pomerance can show that there are infinitely many $n$ for which $P(n) < P(n+1) < P(n+2)$. Are there infinitely many for which $P(n) > P(n+1) > P(n+2)$? They believe that each of these conditions is satisfied by a set of $n$ whose asymptotic density is $\frac{1}{6}$.

P. Erdös and Carl Pomerance, On the largest prime factors of $n$ and $n+1$, *Aequationes Math.* **17** (1978) 311–321.

**B47.** Selfridge notices that $2^2 - 2$ divides $n^2 - n$ for all $n$, that $2^{2^2} - 2^2$ divides $n^{2^2} - n^2$ and $2^{2^{2^2}} - 2^{2^2}$ divides $n^{2^{2^2}} - n^{2^2}$ and asks for what $a$ and $b$ does $2^a - 2^b$ divide $n^a - n^b$ for all $n$?

**B48.** David Silverman noticed that if $p_n$ is the $n$th prime, then

$$\prod_{n=1}^{m} \frac{p_n + 1}{p_n - 1}$$

is an integer for $m = 1, 2, 3, 4$, and 8, and asked is it ever again an integer?

Wagstaff asked for an elementary proof (e.g., without using properties of the Riemann $\zeta$-function) that

$$\prod \frac{p^2 + 1}{p^2 - 1} = \frac{5}{2}$$

where the product is taken over all primes.

# C. Additive Number Theory

**C1.** One of the most infamous problems is Goldbach's conjecture that every even number greater than 4 is expressible as the sum of two odd primes. Vinogradov proved that every *odd* number greater than $3^{3^{15}}$ is the sum of *three* primes and Chen Jing-Run has shown that all large enough even numbers are the sum of a prime and the product of at most two primes.

"Conjecture A" of Hardy and Littlewood (cf. A1, A8) is that the number, $N_2(n)$, of representations of an even number $n$ as the sum of two odd primes, is given asymptotically by

$$N_2(n) \sim \frac{2cn}{(\ln n)^2} \prod \left(\frac{p-1}{p-2}\right),$$

where, as in A8, $2c \approx 1.3203$ and the product is taken over all odd prime divisors of $n$.

Stein and Stein have calculated $N_2(n)$ for $n < 10^5$ and have found values of $n$ for which $N_2(n) = k$ for all $k < 1911$. It is conjectured that $N_2(n)$ takes all positive integer values.

Let $\varphi(n)$ be Euler's totient function (B36) so that if $p$ is prime, $\varphi(p) = p - 1$. If the Goldbach conjecture is true, then there are, for each number $m$, (prime) numbers $p$, $q$, such that

$$\varphi(p) + \varphi(q) = 2m.$$

If we relax the condition that $p$ and $q$ be prime, then it should be easier to show that there are always numbers $p$ and $q$ satisfying this equation. Erdös and Leo Moser ask if this can be done.

Chen Jing-Run, On the representation of a large even number as the sum of a prime and the product of at most two primes, *Sci. Sinica* **16** (1973) 157–176; *MR* **55** #7959; II **21** (1978) 421–430; *MR* **80e**:10037.

J. G. van der Corput, Sur l'hypothèse de Goldbach pour presque tous les nombres pairs, *Acta Arith.* **2** (1937) 266–290.
N. G. Čudakov, On the density of the set of even numbers which are not representable as the sum of two odd primes, *Izv. Akad. Nauk SSSR Ser. Mat.* **2** (1938) 25–40.
N. G. Čudakov, On Goldbach-Vinogradov's theorem, *Ann. of Math.* (2) **48** (1947) 515–545; *MR* **9**, 11.
T. Estermann, On Goldbach's problem: proof that almost all even positive integers are sums of two primes, *Proc. London Math. Soc.* (2) **44** (1938) 307–314.
T. Estermann, Introduction to modern prime number theory, *Cambridge Tracts in Mathematics* **41**, 1952.
H. L. Montgomery and R. C. Vaughan, The exceptional set in Goldbach's problem, *Acta Arith.* **27** (1975) 353–370.
Pan Cheng Dong, Ding Xia Xi and Wang Yuan, On the representation of every large even integer as the sum of a prime and an almost prime, *Sci. Sinica* **18**(1975) 599–610; *MR* **57** #5897.
P. M. Ross, On Chen's theorem that each large even number has the form $p_1 + p_2$ or $p_1 + p_2p_3$, *J. London Math. Soc.* (2) **10** (1975) 500–506.
M. L. Stein and P. R. Stein, New experimental results on the Goldbach conjecture, *Math. Mag.* **38** (1965) 72–80; *MR* **32** #4109.
Robert C. Vaughan, On Goldbach's problem, *Acta Arith.* **22** (1972) 21–48.
Robert C. Vaughan, A new estimate for the exceptional set in Goldbach's problem, *Proc. Sympos. Pure Math. Amer. Math. Soc.* **24** (Analytic Number Theory, St Louis, 1972) 315–319.
I. M. Vinogradov, Representation of an odd number as the sum of three primes, *Dokl. Akad. Nauk SSSR* **15** (1937) 169–172.
I. M. Vinogradov, Some theorems concerning the theory of primes, *Mat. Sb. N.S.* **2** **(44)** (1937) 179–195.
Dan Zwillinger, A Goldbach conjecture using twin primes, *Math. Comput.* **33** (1979) 1071; *MR* **80b**:10071.

**C2.** Let $f(n)$ be the number of ways of representing $n$ as the sum of (one or more) *consecutive* primes. For example $5 = 2 + 3$ and $41 = 11 + 13 + 17 = 2 + 3 + 5 + 7 + 11 + 13$ so that $f(5) = 2$, $f(41) = 3$. Leo Moser has shown that

$$\lim_{x\to\infty} \frac{1}{x} \sum_{n=1}^{x} f(n) = \ln 2$$

and he asks: is $f(n) = 1$ infinitely often? Is $f(n) = k$ solvable for every $k$? Do the numbers for which $f(n) = k$ have a density for every $k$? Is $\limsup f(n) = \infty$?

Erdös asks if there is an infinite sequence of integers $1 < a_1 < a_2 < \cdots$ such that $f(n)$, the number of solutions of $a_i + a_{i+1} + \cdots + a_k = n$, tends to infinity with $n$. He notes that if we insist that $k > i$, then it is not even known if $f(n) > 0$ for all but finitely many $n$. If $a_i = i$, $f(n)$ is the number of odd divisors of $n$.

L. Moser, Notes on number theory III. On the sum of consecutive primes, *Canad. Math. Bull.* **6** (1963) 159–161; *MR* **28** #75.

**C3.** Gardiner and others define *lucky numbers* by modifying the sieve of Eratosthenes in the following way. From the natural numbers strike out all even ones, leaving the odd numbers. Apart from 1, the first remaining number is 3. Strike out every third member (those of form $6k - 1$) in the new sequence, leaving

$$1, 3, 7, 9, 13, 15, 19, 21, 25, 27, 31, 33, \ldots$$

The next number remaining is 7. Strike out every seventh term (numbers $42k - 23$, $42k - 3$) in this sequence. Next 9 remains: strike out every ninth term from what's left, and so on, until we are left with the **lucky numbers**

1, 3, 7, 9, 13, 15, 21, 25, 31, 33, 37, 43, 49, 51, 63, 67, 69, 73, 75, 79, 87, 93, 99, 105, 111, 115, 127, 129, 133, 135, 141, 151, 159, 163, 169, 171, 189, 193, 195, 201, 205, 211, 219, 223, 231, . . . .

Many questions arise concerning lucky numbers, parallel to the classical ones asked about primes. For example, if $L_2(n)$ is the number of solutions of $l + m = n$, where $n$ is even and $l$ and $m$ are lucky, then Stein and Stein find values of $n$ such that $L_2(n) = k$ for all $k \leq 1769$, and there is a corresponding conjecture to that made in C1.

W. E. Briggs, Prime-like sequences generated by a sieve process, *Duke Math. J.* **30** (1963) 297–312, *MR* **26** #6145.

R. G. Buschman and M. C. Wunderlich, Sieve-generated sequences with translated intervals, *Canad. J. Math.* **19** (1967) 559–570; *MR* **35** #2855

R. G. Buschman and M. C. Wunderlich, Sieves with generalized intervals, *Boll. Un. Mat. Ital.* (3) **21** (1966) 362–367.

Paul Erdös and Eri Jabotinsky, On sequences of integers generated by a sieving process. I, II. *Nederl Akad. Wetensch. Proc.* Ser. A **61** = *Indag. Math.* **20** (1958) 115–128; *MR* **21** #2628.

Verna Gardiner, R. Lazarus, N. Metropolis, and S. Ulam, On certain sequences of integers defined by sieves, *Math. Mag.* **29** (1956) 117–122; *MR* **17**, 711.

David Hawkins and W. E. Briggs, The lucky number theorem, *Math. Mag.* **31** (1957–58) 81–84, 277–280; *MR* **21** #2629, 2630.

M. C. Wunderlich, Sieve generated sequences, *Canad. J. Math.* **18** (1966) 291–299; *MR* **32** #5625.

M. C. Wunderlich, A general class of sieve-generated sequences, *Acta Arith.* **16** (1969–70) 41–56; *MR* **39** #6852.

M. C. Wunderlich and W. E. Briggs, Second and third term approximations of sieve-generated sequences, *Illinois J. Math.* **10** (1966) 694–700; *MR* **34** #153.

**C4.** Ulam constructed increasing sequences of positive integers by starting from arbitary $u_1$ and $u_2$ and continuing with those numbers which can be expressed in just one way as the sum of two distinct earlier members of the sequence. Recamán asked some of the questions which arise in connexion with the **U-numbers** ($u_1 = 1$, $u_2 = 2$)

1, 2, 3, 4, 6, 8, 11, 13, 16, 18, 26, 28, 36, 38, 47, 48, 53, 57, 62, 69, 72, 77, 82, 87, 97, 99, 102, 106, 114, 126, 131, 138, 145, 148, 155, 175, 177, 180, 182, 189, 197, 206, 209, 219, . . .

(1) Can the sum of two consecutive U-numbers, apart from $1 + 2 = 3$, be a U-number?

(2) Are there infinitely many numbers

$$23, 25, 33, 35, 43, 45, 67, 92, 94, 96, \ldots$$

which are *not* the sum of two U-numbers?

(3) (Ulam) Do the U-numbers have positive density?

(4) Are there infinitely many pairs,

$$(1, 2), \quad (2, 3), \quad (3, 4), \quad (47, 48), \quad \ldots$$

of consecutive U-numbers?

(5) Are there arbitrarily large gaps in the sequence of U-numbers?

In answer to Question 1 Frank Owens noticed that $u_{19} + u_{20} = 62 + 69 = 131 = u_{31}$. In answer to Question 4 Muller calculated 20000 terms and found no further examples. On the other hand, more than 60% of these terms differed from another by exactly 2. David Zeitlin asks about the behavior of $a(n)$ and $b(n)$, defined by

$$u_{n+3} = u_{n+2-a(n)} + u_{n+1-b(n)} \qquad (a(n) \le b(n), n \ge 0)$$

| $n =$ | 1 | 2 | 3 | 4 | 5 | 6 | 7 | 8 | 9 | 10 | 11 | 12 | 13 | 14 | 15 | 16 |
|---|---|---|---|---|---|---|---|---|---|---|---|---|---|---|---|---|
| $a(n) =$ | 0 | 0 | 0 | 0 | 0 | 0 | 0 | 0 | 0 | 0 | 0 | 1 | 0 | 0 | 0 | 5 |
| $b(n) =$ | 1 | 1 | 2 | 2 | 4 | 4 | 6 | 3 | 8 | 5 | 10 | 6 | 13 | 10 | 12 | 6 |

| $n =$ | 17 | 18 | 19 | 20 | 21 | 22 | 23 | 24 | 25 | 26 | 27 | 28 | 29 | ... |
|---|---|---|---|---|---|---|---|---|---|---|---|---|---|---|
| $a(n) =$ | 2 | 0 | 1 | 2 | 3 | 4 | 0 | 0 | 0 | 0 | 9 | 10 | 4 | ... |
| $b(n) =$ | 9 | 16 | 14 | 13 | 12 | 11 | 22 | 22 | 22 | 21 | 10 | 20 | 17 | ... |

He notes that $b(n) = \varphi(n)$ for $n = 1, 2, 3, 4, 5, 7, 11, 21$, and 23 and asks if this occurs infinitely often. [The strong law of small numbers again?] He also asks if the sequence of U-numbers is *complete* in the sense that every positive number is expressible as the sum of distinct members of the sequence. He asks if the Fibonacci numbers (and the Lucas numbers) are always the sum of at most two U-members.

A. M. Mian and S. Chowla, on the $B_2$-sequence of Sidon, *Proc. Nat. Acad. Sci. India. Sect. A* **14** (1949) 3–4; *MR* **7**, 243.

P. Muller, M.Sc. thesis, Univ. of Buffalo, 1966.

R. Queneau, Sur les suites *s*-additives, *J. Combinatorial Theory* **12** (1972) 31–71.

Bernardo Recamán, Questions on a sequence of Ulam, *Amer. Math. Monthly* **80** (1973) 919–920.

S. M. Ulam, *Problems in Modern Mathematics*, Interscience, N.Y. 1964, ix.

M. C. Wunderlich, The improbable behaviour of Ulam's summation sequence, in *Computers and Number Theory*, Academic Press, 1971, 249–257.

**C5.** Leo Moser asked and Selfridge, Straus, and others largely settled to what extent the sums of all the pairs of numbers in a set determine the set. They show that if the cardinality of the set is not a power of two, then the members are determined. Suppose that $y_1, y_2, \ldots, y_s$ are the sums $x_i + x_j$ $(i \ne j)$ of the numbers $x_1, x_2, \ldots, x_{2^k}$, so that $s = 2^{k-1}(2^k - 1)$. Are there more than two sets $\{x_i\}$ which give rise to the same set $\{y_j\}$? If $k = 3$ there may be three such sets, for example,

$$\{\pm 1, \pm 9, \pm 15, \pm 19\}, \qquad \{\pm 2, \pm 6, \pm 12, \pm 22\}, \qquad \{\pm 3, \pm 7, \pm 13, \pm 21\}$$

but there can't be more than three. For $k > 3$ the problem is open.

The corresponding problem where sums of *triples* of elements of a set are given is also settled, except in two cases: do the sums of three distinct elements of $\{x_1, x_2, \ldots, x_n\}$ determine the set if $n = 27$ or $n = 486$? The corresponding problem for sums of *four* distinct elements was settled by Ewell.

John A. Ewell, On the determination of sets by sets of sums of fixed order, *Canad. J. Math.* **20** (1968) 596–611.
B. Gordon, A. S. Fraenkel, and E. G. Straus, On the determination of sets by the sets of sums of a certain order, *Pacific J. Math.* **12** (1962) 187–196; *MR* **27** #3576.
J. L. Selfridge and E. G. Straus, On the determination of numbers by their sums of a fixed order, *Pacific J. Math.* **8** (1958) 847–856; *MR* **22** #4657.

## C6.

An **addition chain** for $n$ is a sequence $1 = a_0 < a_1 < \cdots < a_r = n$ with each member (after the zeroth) the sum of two earlier (but not necessarily distinct) members. For example

$$1,\ 1+1,\ 2+2,\ 4+2,\ 6+2,\ 8+6 \quad \text{and} \quad 1,\ 1+1,\ 2+2,\ 4+2,\ 4+4,\ 8+6$$

are addition chains for 14 of **length** $r = 5$. The minimal length of an addition chain for $n$ is denoted by $l(n)$.

The main unsolved problem is the Scholz conjecture

$$¿ \qquad l(2^n - 1) \le n - 1 + l(n). \qquad ?$$

It has been proved for $n = 2^a, 2^a + 2^b, 2^a + 2^b + 2^c, 2^a + 2^b + 2^c + 2^d$ by Utz, Gioia et al and Knuth, and demonstrated for $1 \le n \le 18$ and $n = 20, 24, 32$ by Knuth and Thurber. Brauer proved the conjecture for those $n$ for which a shortest chain exists which is a **Brauer chain**, that is, one in which each member uses the previous member as a summand. The second of the examples is not a Brauer chain, because the term $4 + 4$ does not use the summand 6. Such an $n$ is called a **Brauer number**. Hansen proved that there were infinitely many non-Brauer numbers, but also that the Scholz conjecture still holds if $n$ has a shortest chain which is a **Hansen chain**, that is one for which there is a subset $H$ of the members such that each member of the chain uses the largest element of $H$ which is less than the member. The second example is a Hansen chain, with $H = \{1, 2, 4, 8\}$. Knuth gives the example 1, 2, 4, 8, 16, 17, 32, 64, 128, 256, 512, 1024, 1041, 2082, 4164, 8328, 8345, 12509 of a Hansen chain ($H = \{1, 2, 4, 8, 16, 32, 64, 128, 256, 512, 1024, 1041, 2082, 4164, 8328, 8345\}$) for $n = 12509$ which is not a Brauer chain (32 does not use 17) and no such short Brauer chain exists for $n = 12509$.

Are there non-Hansen numbers?

Alfred Brauer, On addition chains, *Bull. Amer. Math. Soc.* **45** (1939) 736–739; *MR* **1**, 40.
Paul Erdös, Remarks on number theory III. On addition chains, *Acta Arith.* **6** (1960) 77–81.
A. A. Gioia and M. V. Subbarao, The Scholz-Brauer problem in addition chains II, *Congressus Numerantium XXII*, Proc. 8th Manitoba Conf. Numerical Math. Comput. 1978, 251–274; *MR* **80i**:10078; *Zbl.* 408.10037.
A. A. Gioia, M. V. Subbarao and M. Sugunamma, The Scholz-Brauer problem in addition chains, *Duke Math. J.* **29** (1962) 481–487; *MR* **25** #3898.
W. Hansen, Zum Scholz-Brauerschen Problem, *J. reine angew. Math.* **202** (1959) 129–136; *MR* **25** #2027.
A. M. Il'in, On additive number chains (Russian), *Problemy Kibernet.* **13** (1965) 245–248.
Donald Knuth, *The Art of Computer Programming*, Vol. 2, Addison-Wesley, Reading, Mass., 1969, 398–422.
Arnold Scholz, Aufgabe 253, *Jber. Deutsch. Math.-Verein.* II **47** (1937) *41–42* (supplement).
K. B. Stolarsky, A lower bound for the Scholz-Brauer problem, *Canad. J. Math.* **21** (1969) 675–683; *MR* **40** #114.
E. G. Straus, Addition chains of vectors, *Amer. Math. Monthly* **71** (1964) 806–808.
Edward G. Thurber, The Scholz-Brauer problem on addition chains, *Pacific J. Math.* **49** (1973) 229–242; *MR* **49** #7233.
Edward G. Thurber, Addition chains and solutions of $l(2n) = l(n)$ and $l(2^n - 1) = n + l(n) - 1$, *Discrete Math.* **16** (1976) 279–289; *MR* **55** #5570; *Zbl.* 346.10032.
W. R. Utz, A note on the Scholz-Brauer problem in addition chains, *Proc. Amer. Math. Soc.* **4** (1953) 462–463; *MR* **14**, 949.
C. T. Wyburn, A note on addition chains, *Proc. Amer. Math. Soc.* **16** (1965) 1134.

**C7.** Given $n$ integers $0 < a_1 < a_2 < \cdots < a_n$ with $(a_1, a_2, \ldots, a_n) = 1$, then $N = \sum_{i=1}^{n} a_i x_i$ has a solution in nonnegative integers $x_i$ if $N$ is large enough. $G(a_1, a_2, \ldots, a_n)$ is the greatest $N$ for which there is no such solution. Sylvester showed that $G(a_1, a_2) = (a_1 - 1)(a_2 - 1) - 1$ and that the number of nonrepresentable numbers is $(a_1 - 1)(a_2 - 1)/2$. Brauer and others showed that $G(a_1, a_2, \ldots, a_n) \le \sum_{i=1}^{n-1} a_{i+1} d_i / d_{i+1}$ where $d_i = (a_1, a_2, \ldots, a_i)$. Roberts and Bateman found the value of $G$ if the $a_i$ are in arithmetic progression. Erdös and Graham showed that $G(a_1, a_2, \ldots, a_n) \le 2a_{n-1}\lfloor a_n/n \rfloor - a_n$ (which is best possible if $n = 2$ and $a_2$ is odd). They define $g(n, t) = \max_{\{a_i\}} G(a_1, a_2, \ldots, a_n)$ where the maximum is taken over all $0 < a_1 < a_2 < \cdots < a_n \le t$ with $(a_1, a_2, \ldots, a_n) = 1$. Their theorem shows that $g(n, t) < 2t^2/n$ and the set $\{x, 2x, \ldots, (n-1)x, x^*\}$ with $x = \lfloor t/(n-1) \rfloor$, $x^* = (n-1)\lfloor t/(n-1) \rfloor - 1$, shows $g(n, t) \ge G(x, \ldots, x^*) \ge t^2/(n-1) - 5t$ for $n \ge 2$. Lewin notes that the sets $\{t/2, t-1, t\}$ or $\{t-2, t-1, t\}$ ($t$ even) and $\{(t-1)/2, t-1, t\}$ ($t$ odd) show that $g(3, t) = \lfloor (t-2)^2/2 \rfloor - 1$.

P. T. Bateman, Remark on a recent note on linear forms, *Amer. Math. Monthly* **65** (1958) 517–518.
E. R. Berlekamp, J. H. Conway, and R. K. Guy, *Winning Ways*, Academic Press, London, 1981, Chap. 18.
Alfred Brauer, On a problem of partitions, *Amer. J. Math.* **64** (1942) 299–312.
A. Brauer and B. M. Seelbinder, On a problem of partitions, II, *ibid.* **76** (1954) 343–346.
A. Brauer and J. E. Shockley, On a problem of Frobenius, *J. reine angew. Math.* **211** (1962) 215–220.

J. S. Byrnes, On a partition problem of Frobenius, *J. Combin. Theory Ser. A* **17** (1974) 162–166; *MR* **50** #234.
P. Erdös and R. L. Graham, On a linear diophantine problem of Frobenius, *Acta Arith.* **21** (1972) 399–408.
B. R. Heap and M. S. Lynn, A graph-theoretic algorithm for the solution of a linear diophantine problem of Frobenius, *Numer. Math.* **6** (1964) 346–354; *MR* **30** #4689.
B. R. Heap and M. S. Lynn, On a linear diophantine problem of Frobenius: An improved algorithm, *ibid* **7** (1964) 226–231; *MR* **31** #1227.
Mordechai Lewin, On a linear diophantine problem, *Bull. London Math. Soc.* **5** (1973) 75–78; *MR* **47** #3311.
N. S. Mendelsohn, A linear diophantine equation with applications to non-negative matrices, *Ann. N.Y. Acad. Sci.* **175** (1970) 287–294.
A. Nijenhuis and H. S. Wilf, Representations of integers by linear forms in non-negative integers, *J. Number Theory* **4** (1972) 98–106.
J. B. Roberts, Note on linear forms, *Proc. Amer. Math. Soc.* **7** (1956) 465–469.
Ö. J. Rödseth, On a linear Diophantine problem of Frobenius, *J. reine angew. Math.* **301** (1978) 171–178.
E. S. Selmer and Ö. Beyer, On the linear diophantine problem of Frobenius in three variables, *J. reine angew. Math.* **301** (1978) 161–170.
J. J. Sylvester, *Math. Quest. Educ. Times* **41** (1884) 21.
Herbert S. Wilf, A circle of lights algorithm for the "money changing problem," *Amer. Math. Monthly* **85** (1978) 562–565.

**C8.** The set of integers $\{2^i: 0 \le i \le k\}$, of cardinality $k+1$, has the sums of all of its $2^{k+1}$ subsets distinct. Erdös has asked for the maximum number, $m$, of positive integers $a_1 < a_2 < \cdots < a_m \le 2^k$, with all sums of subsets distinct. With Leo Moser he has shown that $k + 1 \le m \le k + \frac{1}{2}\log k + 1$ where the logarithm is to base 2. Conway and Guy have given a sequence, $u_0 = 0$, $u_1 = 1$, $u_{n+1} = 2u_n - u_{n-r}$ $(n \ge 1)$ where $r$ is the nearest integer to $\sqrt{2n}$, from which may be derived the set of $k+2$ integers $A = \{a_i = u_{k+2} - u_{k+2-i}: 1 \le i \le k+2\}$. They conjecture that this set has subsets with distinct sums (established by Mike Guy for $k \le 40$). For $k \ge 21$, $u_{k+2} < 2^k$, so $m \ge k+2$ for $k \ge 21$, since once a set with the desired cardinality is found, its cardinality may be increased by doubling the size of each member and adjoining the member 1 (or any odd number). Does $A$ always have subsets with distinct sums? We conjecture that it does, and that in essence it gives the best possible solution to the problem, i.e., that $m = k+2$. Erdös offers \$500.00 for a proof or disproof of $m = k + O(1)$.

V. S. Bludov and V. I. Uberman, A certain sequence of additively distinct numbers (Russian) *Kibernetika* (Kiev) **10** (1974) #5, 111–115; *MR* **53** #332; *Zbl.* 291.10043.
V. S. Bludov and V. I. Uberman, On a sequence of additively differing numbers, *Dopovidi Akad. Nauk. Ukrain. SSR Ser. A* (1974) 483–486, 572; *MR* **50** #7007; *Zbl.* 281.10030.
J. H. Conway and R. K. Guy, Sets of natural numbers with distinct sums, *Notices Amer. Math. Soc.* **15** (1968) 345.
J. H. Conway and R. K. Guy, Solution of a problem of P. Erdös, *Colloq. Math.* **20** (1969) 307.
P. Erdös, Problems and results in additive number theory, *Colloque sur la Théorie des Nombres, Bruxelles*, 1955, Liège and Paris, 1956, 127–137, esp. p. 137.

Hansraj Gupta, Some sequences with distinct sums, *Indian J. Pure Math.* **5** (1974) 1093–1109; *MR* **57** #12440.
B. Lindström, On a combinatorial problem in number theory, *Canad. Math. Bull.* **8** (1965) 477–490.
B. Lindström, Om ett problem av Erdös for talfoljder, *Nordisk Mat. Tidskrift* **16**, 1–2 (1968) 29–30, 80.
Paul Smith, Problem E 2536*, Amer. Math Monthly **82** (1975) 300. Solutions and comments, **83** (1976) 484.
V. I. Uberman, The Conway-Guy conjecture and the density of almost geometric sequences (Russian).
V. I. Uberman, On the theory of a method of determining numbers whose sums do not coincide (Russian), *Proc. Sem. Methods Math. Simulation & Theory Elec. Circuits, Izdat Nauk Dumka* (Kiev) 1973, 76–78, 203; *MR* **51** #7363.
V. I. Uberman, Approximation of additively differing numbers (Russian) *Proc. Sem. Methods Math. Simulation & Theory Elec. Circuits, Naukova Dumka*, Kiev, **11** (1973) 221–229; *MR* **51** #7363; *Zbl.* 309.68048.
V. I. Uberman and V. I. Šleĭnikov, A computer-aided investigation of the density of additive detecting number systems (Russian), *Akad. Nauk Ukrain. SSR, Fiz.-Tehn. Inst. Nizkikh Temperatur*, Kharkov, 1978; 60pp.

**C9.** Suppose that $m$ is the maximum number of integers $1 \le a_1 < a_2 < \cdots < a_m \le n$ for which the sums of pairs, $a_i + a_j$, are all different. It is known that

$$n^{1/2}(1 - \varepsilon) < m \le n^{1/2} + n^{1/4} + 1.$$

The upper bound is due to Lindstrom, improving a result of Erdös and Turán. The lower bound is due to Singer. Erdös and Turán ask, is $m = n^{1/2} + O(1)$? Erdös offers \$500 for settling this equation.

If $\{a_i\}$ continues as an infinite sequence, Erdös and Turán proved that $\limsup a_k/k^2 = \infty$ and gave a sequence with $\liminf a_k/k^2 < \infty$. There is such a sequence with $a_k < ck^3$ for all $k$, and Ajtai, Komlós and Szemerédi have recently shown that $a_k = O(k^3)$ is possible.

Erdös notes that $\sum_{i=1}^{x} a_i^{-1/2} < c(\ln x)^{1/2}$ and asks if this is best possible.

If $f(n)$ is the number of solutions of $n = a_i + a_j$, is there a sequence with $\lim f(n)/\ln n = c$? Erdös and Turán conjecture that if $f(n) > 0$ for all sufficiently large $n$, or if $a_k < ck^2$ for all $k$, then $\limsup f(n) = \infty$; Erdös also offers \$500 for settling this question.

Graham and Sloane rephrase the question in two more obviously packing forms:

Let $v_\alpha(k)$ [respectively $v_\beta(k)$] be the smallest $v$ such that there is a $k$-element set $A = \{0 = a_1 < a_2 < \cdots < a_k\}$ of integers with the property that the sums $a_i + a_j$ for $i < j$ [respectively $i \le j$] belong to $[0, v]$ and represent each element of $[0, v]$ at most once. The set $A$ associated with $v_\beta$ is often called a $B_2$-**sequence** (compare E28).

They give the values of $v_\alpha$ and $v_\beta$ displayed in Table 3 and note that the bounds

$$2k^2 - O(k^{3/2}) < v_\alpha(k),\ v_\beta(k) < 2k^2 + O(k^{36/23})$$

follow from a modification of the Erdös-Turán argument.

Table 3, Values of $v_\alpha(k)$, $v_\beta(k)$ and Exemplary Sets.

| $k$ | $v_\alpha(k)$ | Example of $A$. | $v_\beta(k)$ | Example of $A$. |
|---|---|---|---|---|
| 2 | 1 | $\{0, 1\}$ | 2 | $\{0, 1\}$ |
| 3 | 3 | $\{0, 1, 2\}$ | 6 | $\{0, 1, 3\}$ |
| 4 | 6 | $\{0, 1, 2, 4\}$ | 12 | $\{0, 1, 4, 6\}$ |
| 5 | 11 | $\{0, 1, 2, 4, 7\}$ | 22 | $\{0, 1, 4, 9, 11\}$ |
| 6 | 19 | $\{0, 1, 2, 4, 7, 12\}$ | 34 | $\{0, 1, 4, 10, 12, 17\}$ |
| 7 | 31 | $\{0, 1, 2, 4, 8, 13, 18\}$ | 50 | $\{0, 1, 4, 10, 18, 23, 25\}$ |
| 8 | 43 | $\{0, 1, 2, 4, 8, 14, 19, 24\}$ | 68 | $\{0, 1, 4, 9, 15, 22, 32, 34\}$ |
| 9 | 63 | $\{0, 1, 2, 4, 8, 15, 24, 29, 34\}$ | 88 | $\{0, 1, 5, 12, 25, 27, 35, 41, 44\}$ |
| 10 | 80 | $\{0, 1, 2, 4, 8, 15, 24, 29, 34, 46\}$ | 110 | $\{0, 1, 6, 10, 23, 26, 34, 41, 53, 55\}$ |

M. Ajtai, J. Komlós and E. Szemerédi, *European Combin. J.* (to appear).

R. C. Bose and S. Chowla, Theorems in the additive theory of numbers, *Comment. Math. Helv.* **37** (1962–63) 141–147.

P. Erdös and W. H. J. Fuchs, On a problem of additive number theory, *J. London Math. Soc.* **31** (1956) 67–73.

P. Erdös and E. Szemerédi, The number of solutions of $m = \sum_{i=1}^{k} x_i^k$, *Proc. Symp. Pure Math. Amer. Math. Soc.* **24** (1973) 83–90.

P. Erdös and P. Turán, On a problem of Sidon in additive number theory, and on some related problems, *J. London Math. Soc.* **16** (1941) 212–215; *MR* **3**, 270. Addendum, **19** (1944) 208; *MR* **7**, 242.

R. L. Graham and N. J. A. Sloane, On additive bases and harmonious graphs, *SIAM J. Alg. Discrete Math.* **1** (1980) 382–404.

H. Halberstam and K. F. Roth, *Sequences*, Vol. I, Oxford Univ. Press, 1966, Chapter II.

F. Krückeberg, $B_2$-Folgen und verwandte Zahlenfolgen, *J. reine angew. Math.* **206** (1961) 53–60.

B. Lindstrom, An inequality for $B_2$-sequences, *J. Combin. Theory* **6** (1969) 211–212; *MR* **38** #4436.

J. Singer, A theorem in finite projective geometry and some applications to number theory, *Trans. Amer. Math. Soc.* **43** (1938) 377–385.

Alfred Stöhr, Gelöste und ungelöste Fragen über Basen der natürlichen Zahlenreihe I, II, *J. reine angew. Math.* **194** (1955) 40–65, 111–140; *MR* **17**, 713.

**C10.** Singer's result, mentioned in C9, is based on **perfect difference sets**, i.e., a set of residues $a_1, a_2, \ldots, a_{k+1} \pmod{n}$ such that every nonzero residue (mod $n$) can be expressed uniquely in the form $a_i - a_j$. Perfect difference sets can exist only if $n = k^2 + k + 1$, and Singer proved that such a set exists whenever $k$ is a prime power. Marshall Hall has shown that numerous non-prime-powers *cannot* serve as values of $k$ and Evans and Mann that there is no such $k < 1600$ that is not a prime power. It is conjectured that no perfect difference set exists unless $k$ is a prime power.

Can a given finite sequence, which contains no repeated differences, be extended to form a perfect difference set?

Dean Hickerson asks for the maximum number $r$ such that the integers $1 \le a_1 < a_2 < \cdots < a_r \le n$ have differences $a_j - a_i$, $j > i$ among which the integer $s$ occurs at most $2s$ times.

Graham and Sloane exhibit the problem of difference sets as the modular version of the packing problems of C9. They define $v_\gamma(k)$ [respectively $v_\delta(k)$] as the smallest number $v$ such that there exists a subset $A = \{0 = a_1 < a_2 < \cdots < a_k\}$ of the integers (mod $v$) with the property that each $r$ can be written in at most one way as $r \equiv a_i + a_j$ (mod $v$) with $i < j$ [respectively $i \le j$].

Their interest in $v_\gamma$ is in its application to **error-correcting codes**. If $A(k, 2d, w)$ is the maximum number of binary vectors with $w$ ones and $k - w$ zeros (**words** of **length** $k$ and **weight** $w$) such that any two vectors differ in at least $2d$ places, then (for $d = 3$)

$$A(k, 6, w) \geq \binom{k}{w} \Big/ v_\gamma(k)$$

(and the result for general $d$ uses sets for which all sums of $d - 1$ distinct elements are distinct modulo $v$).

They note that $A(k, 2d, w)$ has been studied by Erdös, Hanani, Schönheim, Stanton, Kalbfleisch, and Mullin in the context of extremal set theory. Let $D(t, k, v)$ be the maximum number of $k$-element subsets of a $v$-element set $S$ such that every $t$-element subset of $S$ is contained in at most one of the $k$-element subsets. Then $D(t, k, v) = A(v, 2k - 2t + 2, k)$.

The values of $v_\delta$ in Table 4 are from Baumert's Table 6.1 and those of $v_\gamma$ from Graham and Sloane who give the following bounds

$$k^2 - O(k) < v_\gamma(k) < k^2 + O(k^{36/23}),$$
$$k^2 - k + 1 \leq v_\delta(k) < k^2 + O(k^{36/23}).$$

Equality holds on the left of this last whenever $k - 1$ is a prime power.

Table 4. Values of $v_\gamma(k)$, $v_\delta(k)$ and Exemplary Sets.

| $k$ | $v_\gamma(k)$ | Example of $A$ | $v_\delta(k)$ | Example of $A$ |
|---|---|---|---|---|
| 2 | 2 | {0, 1} | 3 | {0, 1} |
| 3 | 3 | {0, 1, 2} | 7 | {0, 1, 3} |
| 4 | 6 | {0, 1, 2, 4} | 13 | {0, 1, 3, 9} |
| 5 | 11 | {0, 1, 2, 4, 7} | 21 | {0, 1, 4, 14, 16} |
| 6 | 19 | {0, 1, 2, 4, 7, 12} | 31 | {0, 1, 3, 8, 12, 18} |
| 7 | 28 | {0, 1, 2, 4, 8, 15, 20} | 48 | {0, 1, 3, 15, 20, 38, 42} |
| 8 | 40 | {0, 1, 5, 7, 9, 20, 23, 35} | 57 | {0, 1, 3, 13, 32, 36, 43, 52} |
| 9 | 56 | {0, 1, 2, 4, 7, 13, 24, 32, 42} | 73 | {0, 1, 3, 7, 15, 31, 36, 54, 63} |
| 10 | 72 | {0, 1, 2, 4, 7, 13, 23, 31, 39, 59} | 91 | {0, 1, 3, 9, 27, 49, 56, 61, 77, 81} |

L. D. Baumert, *Cyclic Difference Sets*, Lecture notes in Math. **182**, Springer-Verlag, New York, 1971.

M. R. Best, A. E. Brouwer, F. J. MacWilliams, A. M. Odlyzko, and N. J. A. Sloane, Bounds for binary codes of length less than 25, *IEEE Trans. Information Theory* **IT-24** (1978) 81–93.

P. Erdös and H. Hanani, On a limit theorem in combinatorical analysis, *Publ. Math. Debrecen* **10** (1963) 10–13.

T. A. Evans and H. Mann, On simple difference sets, *Sankhyā* **11** (1951) 357–364; *MR* **13**, 899.
R. L. Graham and N. J. A. Sloane, Lower bounds for constant weight codes, *IEEE Trans. Information Theory* **IT-26** (1980).
R. L. Graham and N. J. A. Sloane, On constant weight codes and harmonious graphs, *Utilitas Math.* (to appear).
J. I. Hall, A. J. E. M. Jensen, A. W. J. Kolen and J. H. van Lint, Equidistant codes with distance 12, *Discrete Math.* **17** (1977) 71–83.
M. Hall, Cyclic projective planes, *Duke Math. J.* **14** (1947) 1079–1090; *MR* **9**, 370.
D. McCarthy, R. C. Mullin, P. J. Schellenberg, R. G. Stanton, and S. A. Vanstone, On approximations to a projective plane of order 6, *Ars Combinatoria* **2** (1976) 111–168.
F. J. MacWilliams and N. J. A. Sloane, *The Theory of Error-Correcting Codes*, North-Holland, 1977.
J. Schönheim, On maximal systems of $k$-tuples, *Stud. Sci. Math. Hungar.* **1** (1966) 363–368.
R. G. Stanton, J. G. Kalbfleisch, and R. C. Mullin, Covering and packing designs, in *Proc. 2nd Conf. Combin. Math. and Appl.*, Chapel Hill, 1970, 428–450.

**C11.** Bose and Chowla obtained an analogous result to that of C9 when the subsets of *three* distinct elements all have different sums. If $m_3$ is the largest number of $a_i$, $1 \le a_1 < a_2 < \cdots < a_{m_3} \le n$ such that the sums $a_i + a_j + a_k$ are all distinct, then they showed that $m_3 \ge n^{1/3}(1 + o(1))$, and they asked if $m_3 \ge (1 + \varepsilon)n^{1/3}$.

Lindström has shown that if $m_4$ is the maximum number of integers $\le n$ whose sets of *four* have distinct sums, then $m_4 < (8n)^{1/4} + O(n^{1/8})$.

S. C. Bose and S. Chowla, *Report Inst. Theory of Numbers*, Univ. of Colorado, Boulder, 1959.
B. Lindström, A remark on $B_4$-sequences, *J. Combin. Theory* **7** (1969) 276–277.

**C12.** The covering problem which is dual to the packing problem C9 is due to Rohrbach. A sequence $A$ is called an **additive basis of order *h*** or ***h*-basis** (for $n$) if every nonnegative integer, not greater than $n$, is expressible as the sum of at most $h$, not necessarily distinct, members of $A$. If $h$ is the *least* number for which $A$ has this property, $h$ is called the **exact order** of $A$. If $k_h(n)$ is the least number of elements in an $h$-basis for $n$, Rohrbach showed that $k_h(n) < hn^{1/h}$ and that $\frac{1}{2}k_2^2(1 - 0.0016) > n$, a result successively improved by Leo Moser, Riddell, and Klotz by replacing 0.0016 first by 0.0194, then 0.0269, and then 0.0369. They also proved, for $h \ge 3$, $0 < \varepsilon < 1$, and $n$ sufficiently large, that

$$\frac{k^h}{h!}\left\{1 - \frac{(1-\varepsilon)\cos \pi/h}{2 + \cos \pi/h}\right\} > n.$$

Let $n(h, k)$ be the largest integer for which an $h$-basis of $k$ elements exists. The problem of determining $n(h, k)$ has appeared very frequently, often as a **postage stamp** (or coin) **problem**: see the paper of Alter and Barnett for more detail and a very extensive bibliography.

Rohrbach proved $n(2,k) \geq k^2/4$ and conjectured that $n(2,k)/k^2 \to \frac{1}{4}$ as $k \to \infty$ but this was shown to be false by Hämmerer and Hofmeister. A basis for $n(2,k)$ with $k_2(n)$ elements is called **extremal**. If the elements of a 2-basis for $n$ do not exceed $n/2$, the basis is called **restricted**. Rohrbach's bases are **symmetric** about $n/4$ (and hence restricted). For example $n(2,10) = 46$, but the only extremal bases for 46,

$$1, 2, 3 \text{ or } 5, 7, 11, 15, 19, 21, 22, 24$$

are restricted. Some care is needed in reading the literature: this result may appear as $n(2,11) = 46$ in contexts where every nonnegative integer is to be expressed as the sum of *exactly* $h$ members, not necessarily distinct, of the basis. This corresponds to the inclusion of a zero denomination stamp or coin. If an extremal basis is restricted, is it necessarily symmetric?

Stöhr and several later writers showed that

$$n(h,2) = \lfloor (h^2 + 6h + 1)/4 \rfloor,$$

and Hofmeister that, for $h \geq 34$,

$$\tfrac{4}{81}h^3 + \tfrac{2}{3}h^2 + \tfrac{66}{27}h \leq n(h,3) \leq \tfrac{4}{81}h^3 + \tfrac{2}{3}h^2 + \tfrac{71}{27}h - \tfrac{1}{81}.$$

The present writer suggests that for $h$ large enough, $n(h,k)$ is given by a finite set of polynomials of degree $k$ in $h$. In particular that, for $k = 3$ and $h \geq 20$,

$$¿ \qquad n(h,3) = (4h^3 + 54h^2 + (204 + 3c_r)h + d_r)/81 \qquad ?$$

where $c_r$, $d_r$ are, for $h \equiv r \pmod 9$, given by

| $r =$ | $-4$ | $-3$ | $-2$ | $-1$ | $0$ | $1$ | $2$ | $3$ | $4$ |
|---|---|---|---|---|---|---|---|---|---|
| $c_r =$ | $0$ | $1$ | $3$ | $0$ | $-2$ | $0$ | $3$ | $1$ | $0$ |
| $d_r =$ | $46$ | $-81$ | $-1$ | $-170$ | $0$ | $62$ | $-26$ | $0$ | $-154$ |

For $k$ fixed and greater than 2, and $h$ large, Hofmeister gives the bounds

$$2^{\lfloor k/4 \rfloor}(4/3)^{\lfloor k-4\lfloor k/4 \rfloor \rfloor}(h/k)^k + O(h^{k-1}) \leq n(h,k) \leq h^k/k! + O(h^{k-1}).$$

Graham and Sloane (compare C9, C10) define $n_\alpha(k)$ [respectively $n_\beta(k)$] as the largest number $n$ such that there is a $k$-element set $A = \{0 = a_1 < a_2 < \cdots < a_k\}$ of integers with the property that each $r$ in $[1,n]$ can be written in at least one way as $r = a_i + a_j$ with $i < j$ [respectively $i \leq j$], so that their $n_\beta(k)$ is here written $n(2, k-1)$ (note the "zero proviso" mentioned above), and their $n_\alpha(k)$ corresponds to the problem of two stamps of different denominations, with a zero denomination included.

They give the values for $n_\alpha$ and $n_\beta$ in Table 5. The bounds

$$\tfrac{5}{18}(k-1)^2 < n_\alpha(k),\, n_\beta(k) < 0.4802k^2 + O(k)$$

are essentially due to Hämmerer and Hofmeister and to Klotz.

Table 5. Values of $n_\alpha$ and $n_\beta$ and Exemplary Sets.

| $k$ | $n_\alpha(k)$ | Example of $A$ | $n_\beta(k)$ | Example of $A$ |
|---|---|---|---|---|
| 2 | 1 | {0,1} | 2 | {0,1} |
| 3 | 3 | {0,1,2} | 4 | {0,1,2} |
| 4 | 6 | {0,1,2,4} | 8 | {0,1,3,4} |
| 5 | 9 | {0,1,2,3,6} | 12 | {0,1,3,5,6} |
| 6 | 13 | {0,1,2,3,6,10} | 16 | {0,1,3,5,7,8} |
| 7 | 17 | {0,1,2,3,4,8,13} | 20 | {0,1,2,5,8,9,10} |
| 8 | 22 | {0,1,2,3,4,8,13,18} | 26 | {0,1,2,5,8,11,12,13} |
| 9 | 27 | {0,1,2,3,4,5,10,16,22} | 32 | {0,1,2,5,8,11,14,15,16} |
| 10 | 33 | {0,1,2,3,4,5,10,16,22,28} | 40 | {0,1,3,4,9,11,16,17,19,20} |
| 11 | 40 | {0,1,2,4,5,6,10,13,20,27,34} | 46 | {0,1,2,3,7,11,15,19,21,22,24} |
| 12 | 47 | {0,1,2,3,6,10,14,18,21,22,23,24} | 54 | {0,1,2,3,7,11,15,19,23,25,26,28} |
| 13 | 56 | {0,1,2,4,6,7,12,14,17,21,30,39,48} | 64 | {0,1,3,4,9,11,16,21,23,28,29,31,32} |
| 14 | 65 | {0,1,2,4,6,7,12,14,17,21,30,39,48,57} | 72 | {0,1,3,4,9,11,16,20,25,27,32,33,35,36} |

R. Alter and J. A. Barnett, Remarks on the postage stamp problem with applications to computers, *Congressus Numerantium XIX*, Proc. 8th S.E. Conf. Combin., Graph Theory, Comput., Utilitas Math. 1977, 43–59; *MR* **57** #12246.

R. Alter and J. A. Barnett, A postage stamp problem, *Amer. Math Monthly* **87** (1980) 206–210.

N. Hämmerer and G. Hofmeister, Zu einer Vermutung von Rohrbach, *J. reine angew. Math.* **286/287** (1976) 239–247; *MR* **54** #10181.

E. Härtter, Basen für Gitterpunktmengen, *J. reine angew. Math.* **202** (1959) 153–170; *MR* **22A** #31.

R. L. Heimer and H. Langenbach, The stamp problem, *J. Recreational Math.* **7** (1974) 235–250.

G. Hofmeister, Asymptotische Abschätzungen für dreielementige Extremalbasen in natürlichen Zahlen, *J. reine angew. Math* **232** (1968) 77–101; *MR* **38** #1068.

G. Hofmeister, Endliche additive Zahlentheorie, Kapitel I, Das Reichweitenproblem, Joh. Guttenberg-Univ. Mainz, 1976

G. Hofmeister and H. Schell, Reichweiten von Mengen natürlicher Zahlen I, *Norske Vid. Selsk. Skr.* (Trondheim) **1970** #10, 5pp; *MR* **44** #1642.

W. Klotz, Eine obere Schranke für die Reichweite einer Extremalbasis zweiter Ordnung, *J. reine angew. Math.* **238** (1969) 161–168 (and see 194–220); *MR* **40** #117, 116.

W. F. Lunnon, A postage stamp problem, *Comput. J.* **12** (1969) 377–380; *MR* **40** #6745.

L. Moser, On the representation of 1, 2, . . . , $n$ by sums, *Acta Arith.* **6** (1960) 11–13; *MR* **23A** #133.

L. Moser, J. R. Pounder and J. Riddell, On the cardinality of $h$-bases for $n$, *J. London Math. Soc.* **44** (1969) 397–407; *MR* **39** #162.

L. Moser and J. Riddell. On additive $h$-bases for $n$, *Colloq. Math.* **9** (1962) 287–290; *MR* **26** #1295.

Arnulf Mrose, Untere Schranken für die Reichweiten von Extremalbasen fester Ordnung, *Abh. Math. Sem. Univ. Hamburg* **48** (1979) 118–124; *MR* **80g**:10058.

M. B. Nathanson, Additive $h$-bases for lattice points, 2nd Internat. Conf. Combin. Math., *Annals N.Y. Acad. Sci.* **319** (1979) 413–414; *MR* **81e**: 10041.

J. Riddell, On bases for sets of integers, Master's thesis, Univ. of Alberta, 1960.

J. Riddell and C. Chan, Some extremal 2-bases, *Math. Comput.* **32** (1978) 630–634; *MR* **57** #16244; *Zbl.* 388.10032.

H. Rohrbach, Ein Beitrag zur additiven Zahlentheorie, *Math. Z.* **42** (1937) 1–30; *Zbl.* **15**, 200.

H. Rohrbach, Anwendung eines Satzes der additiven Zahlentheorie auf eine graphentheoretische Frage, *Math. Z.* **42** (1937) 538–542.

R. G. Stanton, J. A. Bate and R. C. Mullin, Some tables for the postage stamp problem, *Congressus Numerantium XII*, Proc. 4th Manitoba Conf. Numer. Math. Winnipeg 1974, 351–356; *MR* **51** #7887.

**C13.** Just as C10 was the modular version of the packing problem C9, so we can ask the modular version of the corresponding covering problem C12.

Graham and Sloane complete their octad of definitions with $n_\gamma(k)$ [respectively $n_\delta(k)$] as the largest number $n$ such that there is a subset $A = \{0 = a_1 < a_2 < \cdots < a_k\}$ of the residue classes modulo $n$ with the property that each $r$ can be written in at least one way as $r \equiv a_i + a_j \pmod n$ with $i < j$ [respectively $i \le j$].

Table 6. Values of $n_\gamma$ and $n_\delta$ and Exemplary Sets.

| $k$ | $n_\gamma(k)$ | Example of $A$ | $n_\delta(k)$ | Example of $A$ |
|---|---|---|---|---|
| 2 | 1 | — | 3 | $\{0,1\}$ |
| 3 | 3 | $\{0,1,2\}$ | 5 | $\{0,1,2\}$ |
| 4 | 6 | $\{0,1,2,4\}$ | 9 | $\{0,1,3,4\}$ |
| 5 | 9 | $\{0,1,2,4,7\}$ | 13 | $\{0,1,2,6,9\}$ |
| 6 | 13 | $\{0,1,2,3,6,10\}$ | 19 | $\{0,1,3,12,14,15\}$ |
| 7 | 17 | $\{0,1,2,3,4,8,13\}$ | 21 | $\{0,1,2,3,4,10,15\}$ |
| 8 | 24 | $\{0,1,2,4,8,13,18,22\}$ | 30 | $\{0,1,3,9,11,12,16,26\}$ |
| 9 | 30 | $\{0,1,2,4,10,15,17,22,28\}$ | 35 | $\{0,1,2,7,8,11,26,29,30\}$ |
| 10 | 36 | $\{0,1,2,3,6,12,19,20,27,33\}$ | | |

They give the values in Table 6 and the bounds

$$\tfrac{5}{18}(k-1)^2 < n_\gamma(k),\ n_\delta(k) < \tfrac{1}{2}k^2 + O(k).$$

They call a connected graph with $v$ vertices and $e \ge v$ edges **harmonious** if there is a labelling of the vertices $x$ with distinct labels $l(x)$ so that when an edge $xy$ is labelled with $l(x) + l(y)$, the edge labels form a complete system of residues (mod $e$). Trees (for which $e = v - 1$) are also called harmonious if just one vertex label is duplicated and the edge labels form a complete system (mod $v - 1$). The connexion with the present problem is that $n_\gamma(v)$ is the greatest number of edges in any harmonious graph on $v$ vertices. For example, from Table 6 we note that $n_\gamma(5) = 9$ is attained by the set $\{0,1,2,4,7\}$ so that a maximum of 9 edges can occur in a harmonious graph on 5 vertices (Figure 6).

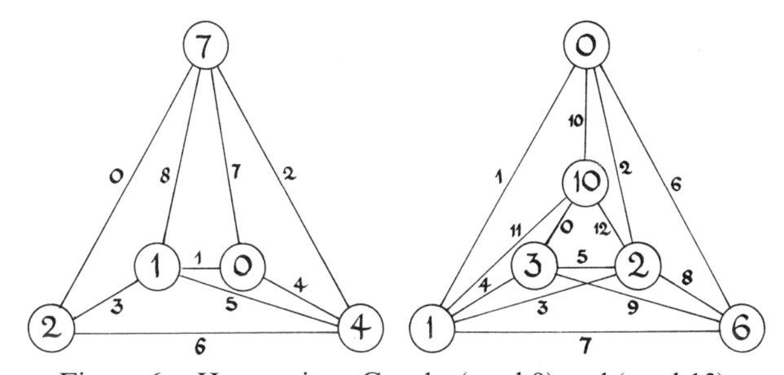


Figure 6. Harmonious Graphs (mod 9) and (mod 13).

Graham and Sloane compare and contrast harmonious graphs with graceful graphs, which will be discussed in the graph theory chapter of a later volume in this series. A graph is **graceful** if, when the vertex labels are chosen from $[0, e]$ and the edge labels calculated by $|l(x) - l(y)|$, the latter are all distinct (i.e., take the values $[1, e]$).

Trees are conjectured to be both harmonious and graceful, but these are open questions. A cycle $C_n$ is harmonious just if $n$ is odd, and graceful just if $n \equiv 0$ or 3 (mod 4). The friendship graph or windmill is harmonious just if $n \not\equiv 2$ (mod 4) and graceful just if $n \equiv 0$ or 1 (mod 4). Fans and wheels are both harmonious and graceful, as is the Petersen graph. The graphs of the five Platonic solids are naturally graceful and one would expect them to be harmonious, but this is not so for the cube, nor for the octahedron.

## C14. **Maximal sum-free sets.**

Denote by $l(n)$ the largest $l$ so that if $a_1, a_2, \ldots, a_n$ are any $n$ distinct natural numbers, one can always find $l$ of them so that $a_{i_j} + a_{i_k} \neq a_m$ for $1 \le j < k \le l$, $1 \le m \le n$ [$j \neq k$, else the set $\{a_i = 2^i | 1 \le i \le n\}$ would imply that $l(n) = 0$]. A remark of Klarner shows that $l(n) > c \ln n$. On the other hand, the set $\{2^i + 0, \pm 1 | 1 < i \le s + 1\}$ implies that $l(3s) < s + 3$, so $l(n) < \frac{1}{3}n + 3$. Selfridge extends this by using the set $\{(3m + t)2^{m-i} | -i < t < i, 1 \le i \le m\}$ to show that $l(m^2) < 2m$. Choi, using sieve methods, has further improved this to $l(n) \ll n^{0.4+\varepsilon}$.

The problem can be generalized to ask if, for every $l$, there is an $n_0 = n_0(l)$ so that if $n > n_0$ and $a_1, a_2, \ldots, a_n$ are any $n$ elements of a group with no product $a_{i_1}a_{i_2} = e$, the unit [here $i_1, i_2$ may be equal, so there is no $a_i$ of order 1 or 2, and no $a_i$ whose inverse is also an $a_i$] then there are $l$ of the $a_i$ such that $a_{i_j}a_{i_k} \neq a_m$, $1 \le j < k \le l$, $1 \le m \le n$. This has not even been proved for $l = 3$.


S. L. G. Choi, On sequences not containing a large sum-free subsequence, *Proc. Amer. Math. Soc.* **41** (1973) 415–418; *MR* **48** #3910.

S. L. G. Choi, On a combinatorial problem in number theory, *Proc. London Math. Soc.* (3) **23** (1971) 629–642; *MR* **45** #1867.

P. H. Diananda and H.-P. Yap, Maximal sum-free sets of elements of finite groups, *Proc. Japan Acad.* **45** (1969) 1–5; *MR* **39** #6968.

Leo Moser, Advanced problem 4317, *Amer. Math Monthly* **55** (1948) 586; solution Robert Steinberg **57** (1950) 345.

Anne Penfold Street, A maximal sum-free set in $A_5$, *Utilitas Math.* **5** (1974) 85–91; *MR* **49** #7156.

Anne Penfold Street, Maximal sum-free sets in abelian groups of order divisible by three, *Bull. Austral. Math. Soc.* **6** (1972) 439–441; *MR* **46** #5484. Corrigenda **7** (1972) 317–318; *MR* **47** #5147.

P. Varnavides, On certain sets of positive density, *J. London Math. Soc.* **34** (1959) 358–560; *MR* **21** #5595.

H. P. Yap, Maximal sum-free sets in finite abelian groups, *Bull. Austral. Math. Soc.* **4** (1971) 217–223; *MR* **43** #2081 [and see *ibid* **5** (1971) 43–54; *MR* **45** #3574; *Nanta Math.* **2** (1968) 68–71; *MR* **38** #3345; *Canad. J. Math.* **22** (1970) 1185–1195; *MR* **42** #1897; *J. Number Theory* **5** (1973) 293–300; *MR* **48** #11356].

**C15.** Erdös and Heilbronn asked for the largest number $k = k(m)$ of distinct residue classes, modulo $m$, so that no subset has sum zero. For example, the set

$$1, -2, 3, 4, 5, 6$$

shows that $k(20) \geq 6$, and in fact equality holds. The pattern of this example shows that

$$k \geq [(-1 + \sqrt{8m + 9})/2] \qquad (m \geq 5).$$

Equality holds for $5 \leq m \leq 24$. However, Selfridge observes that if $m$ is of the form $2(l^2 + l + 1)$, the set

$$1, 2, \ldots, l-1, l, \tfrac{1}{2}m, \tfrac{1}{2}m + 1, \ldots, \tfrac{1}{2}m + l$$

implies that

$$k \geq 2l + 1 = \sqrt{2m - 3}.$$

In fact he conjectures that for any even $m$, this set or the set with $l$ deleted, always gives the best result. For example, $k(42) \geq 9$.

On the other hand, if $p$ is a prime in the interval

$$\tfrac{1}{2}k(k+1) < p < \tfrac{1}{2}(k+1)(k+2),$$

he conjectures that $k(p) = k$, where the set can be simply

$$1, 2, \ldots, k.$$

The case $k(43) = 8$ was confirmed by Clement Lam, so $k$ is not a monotonic function of $m$.

The only case where a better inequality is known than $k \geq \lfloor\sqrt{2m - 3}\rfloor$ is $k(25) \geq \sqrt{50 - 1} = 7$, as is shown by the set 1, 6, 11, 16, 21, 5, 10. If $m$ is of the form $25l(l+1)/2$ and *odd*, then it is possible to improve on the set $1, -2, 3, 4, \ldots$, but if $m$ is of that form and *even*, then the construction already given for $m$ even is always better.

Is $k = \lfloor(-1 + \sqrt{8m + 9})/2\rfloor$ for an infinity of values of $m$?

For which values of $m$ are there realizing sets none of whose members are prime to $m$? For example, $m = 12$: $\{3, 4, 6, 10\}$ or $\{4, 6, 9, 10\}$. Is there a value of $m$ for which *all* realizing sets are of this type?

Erdös and Heilbronn proved that if $a_1, a_2, \ldots, a_k$, $k \geq 3(6p)^{1/2}$, are distinct residues (mod $p$), where $p$ is prime, then every residue (mod $p$) can be written in the form $\sum_{i=1}^{k} \varepsilon_i a_i$, $\varepsilon_i = 0$ or 1. They conjectured that the same holds for $k > 2\sqrt{p}$ and that this is best possible, and Olsen proved this. They further conjectured that the number of distinct residues of the form $a_i + a_j$, $1 \leq i \leq j \leq k$, is at least $2k - 3$; this problem is still open.

P. Erdös, Some problems in number theory, in *Computers in Number Theory*, Academic Press, London and New York, 1971, 405–413.

P. Erdös and H. Heilbronn, On the addition of residue classes mod $p$, *Acta Arith.* **9** (1969) 149–159.

Henry B. Mann and John E. Olsen, Sums of sets in the elementary abelian group of type $(p, p)$, *J. Combin. Theory* **2** (1967) 275–284.
John E. Olsen, An addition theorem, modulo $p$, *J. Combin. Theory* **5** (1968) 45–52.
John E. Olsen, An addition theorem for the elementary abelian group, *J. Combin. Theory* **5** (1968) 53–58.
C. Ryavec, The addition of residue classes modulo $n$, *Pacific J. Math.* **26** (1968) 367–373.
E. Szemerédi, On a conjecture of Erdös and Heilbronn, *Acta Arith.* **17** (1970–71) 227–229.

## C16.

A **nonaveraging set** $A$ of integers $0 \le a_1 < a_2 < \cdots < a_n \le x$ was defined by Erdös and Straus by the property that no $a_i$ shall be the arithmetic mean of any subset of $A$ with more than one element. Denote by $f(x)$ the maximum number of elements in such a set, and by $g(x)$ the maximum number of elements in a subset $B$ of the integers $[0, x]$ such that no two distinct subsets of $B$ have the same arithmetic mean, and by $h(x)$ the corresponding maximum where the subsets of $B$ have different cardinality. Abbott and Erdös and Straus show (by using Szemerédi's result; see E10) that

$$\tfrac{1}{10}\log x + O(1) < \log f(x) < \tfrac{2}{3}\log x + O(1)$$
$$\tfrac{1}{2}\log x - 1 < g(x) < \log x + O(\ln\ln x)$$
$$\sqrt{\log x} - 1 + O(1/\sqrt{\ln x}) < \log h(x) < 2\log\ln x + O(1)$$

and conjecture that $f(x) = \exp(c\sqrt{\ln x}) = o(x^\varepsilon)$ and that $h(x) = (1 + o(1))\log x$. [$\log x = (\ln x)/(\ln 2)$ is the logarithm to the base 2.]

Erdös originally asked for the maximum number, $k(x)$, of integers in $[0, x]$ so that no one divides the sum of any others. Such **nondividing sets** are obviously nonaveraging, so $k(x) \le f(x)$. Straus showed that $k(x) \ge \max\{f(x/f(x)), f(\sqrt{x})\}$.

Abbott has recently shown that if $l(n)$ is the largest $m$ such that *every* set of $n$ integers contains a nonaveraging subset of size $m$, then $l(n) < n^{1/13-\varepsilon}$.

Compare problems C14–16 with E10–14.

H. L. Abbott, On a conjecture of Erdös and Straus on non-averaging sets of integers, *Congressus Numerantium XV, Proc. 5th Brit. Combin. Conf. Aberdeen, 1975, 1–4.*
H. L. Abbott, Extremal problems on non-averaging and non-dividing sets, *Pacific J. Math.* **91** (1980) 1–12.
P. Erdös and E. G. Straus, Non-averaging sets II, in *Combinatorial Theory and its Applications* II, Colloq. Math. Soc. János Bolyai **4**, North-Holland, 1970, 405–411.
E. G. Straus, Non-averaging Sets, *Proc. Symp. Pure Math.* **19** Amer. Math. Soc., Providence 1971, 215–222.

## C17.

**The minimum overlap problem.** Let $\{a_i\}$ be an arbitrary set of $n$ distinct integers, $1 \le a_i \le 2n$, and $\{b_j\}$ be the complementary set $1 \le b_j \le 2n$, with $b_j \ne a_i$. $M_k$ is the number of solutions of $a_i - b_j = k$ $(-2n < k < 2n)$ and $M = \min\max_k M_k$, where the minimum is taken over all sequences $\{a_i\}$. Erdös proved that $M > n/4$; Scherk improved this to $M > (1 - 2^{-1/2})n$ and Swierczkowski to $M > (4 - \sqrt{6})n/5$. Leo Moser obtained the further improvements $M > \sqrt{2}(n-1)/4$ and $M > \sqrt{4-\sqrt{15}}(n-1)$. In the other direction, Motzkin et al obtained examples to show that $M < 2n/5$, contrary to Erdös's conjecture that $M = \frac{1}{2}n$. Is there a number $c$ such that $M \sim cn$?

Leo Moser asks the corresponding question where the cardinality of $\{a_i\}$ is not $n$, but $k$, where $k = \lfloor \alpha n \rfloor$ for some real $\alpha$, $0 < \alpha < 1$.


P. Erdös, Some remarks on number theory (Hebrew, English summary), *Riveon Lematematika*, **9** (1955) 45–48; *MR* **17**, 460.

L. Moser, On the minimum overlap problem of Erdös, *Acta Arith.* **5** (1959) 117–119; *MR* **21** #5594.

T. S. Motzkin, K. E. Ralston, and J. L. Selfridge, Minimum overlappings under translation, *Bull. Amer. Math. Soc.* **62** (1956) 558.

S. Swierczkowski, On the intersection of a linear set with the translation of its complement, *Colloq. Math.* **5** (1958) 185–197; *MR* **21** #1955.


**C18.** Paul Berman asked what is the minimum number of Queens which can be placed on an $n \times n$ chessboard so that no Queen is guarding any other Queen and so that the entire board is guarded by all the Queens? Claude Berge asked for the smallest number of Queens on an ordinary chessboard so that every square is guarded by at least one Queen. He noted that, in graph theory language, this is the same as finding the minimum externally stable set for a graph on 64 vertices with two vertices joined just if they are on the same rank, file, or diagonal. In his notation $\beta = 5$ for Queens (Figure 7a), 8 for Bishops (Figure 7b), and 12 for Knights (Figure 7c). Although here there is no condition that a piece may not guard another piece, the condition is satisfied by the Queens and Bishops, but not by the Knights. Since in Chess a piece does not guard the square it stands on, there are in fact two sets of problems.

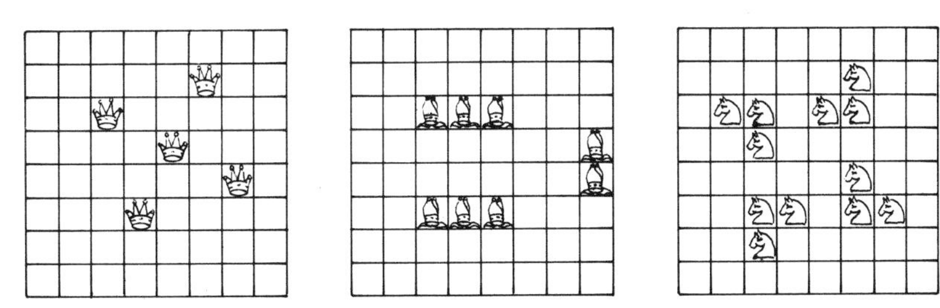

Figure 7. Minimum Covers of the Chessboard by Queens, Bishops and Knights.

For the Queens, Kraitchik gave the following experimental table (in which the entries for $n = 5$ and 6 have been corrected) for an $n \times n$ board.

| $n$ | 5 | 6 | 7 | 8 | 9 | 10 | 11 | 12 | 13 | 14 | 15 | 16 | 17 |
|---|---|---|---|---|---|---|---|---|---|---|---|---|---|
| number of Queens | 3 | 4 | 5 | 5 | 5 | 5 | 5 | 6 | 7 | 8 | 9 | 9 | 9 |

Corresponding configurations for $n = 5, 6, 11$ are shown in Figure 8.

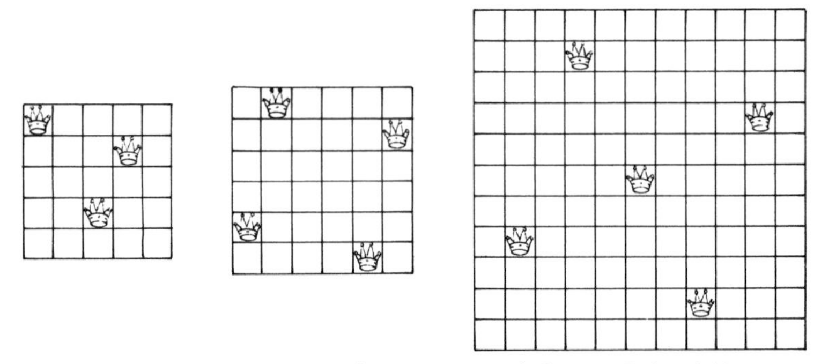

Figure 8. Queens Covering $n \times n$ Boards for $n = 5, 6$ and 11.

If we try to partition the numbers from 1 to $2n$ into $n$ pairs $a_i$, $b_i$ so that the $2n$ numbers $a_i \pm b_i$ fall one in each residue class modulo $2n$, then it is found to be impossible. Less restrictively, Shen and Shen asked that the $2n$ numbers $a_i \pm b_i$ be distinct. They gave examples for $n=3$: 1, 5; 2, 3; 4, 6; for $n = 6$: 1, 10; 2, 6; 3, 9; 4, 11; 5, 8; 7, 12; and $n = 8$: 1, 10; 2, 14; 3, 16; 4, 11; 5, 9; 6, 12; 7, 15; 8, 13; and Selfridge showed that there was always a solution for $n \geq 3$. How many solutions are there for each $n$?

If the condition $b_i = i\ (1 \leq i \leq n)$ is added, we have the reflecting Queens problem: place $n$ Queens on an $n \times n$ chessboard so that no two are on the same rank, file or diagonal, where, on a diagonal, we include reflexions in a mirror in the centre of the zero column (Fig. 9).

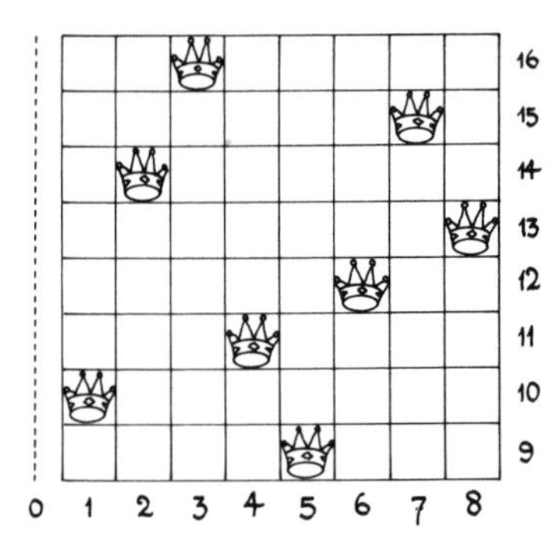


Figure 9. A Solution of the Reflecting Queens Problem.

We can again ask for the number of solutions for each $n$, both in the case where we distinguish between solutions obtained by rotation and reflexion, and where we do not.

Claude Berge, *Theory of Graphs*, Methuen, 1962, p. 41.
Paul Berman, Problem 122, *Pi Mu Epsilon J.* **3** (1959–64) 118, 412.
A. Bruen and R. Dixon, The $n$-queens problem, *Discrete Math.* **12** (1975) 393–395.

B. Hansche and W. Vucenic, On the $n$-queens problem, *Notices Amer. Math. Soc.* **20** (1973) A-568.
G. B. Huff, On pairings of the first $2n$ natural numbers, *Acta Arith.* **23** (1973) 117–126.
D. A. Klarner, The problem of the reflecting queens, *Amer. Math. Monthly* **74** (1967) 953–955; *MR* **40** #7123.
Maurice Kraitchik, *Mathematical Recreations*, Norton, New York 1942, 247–256.
J. D. Sebastian, Some computer solutions to the reflecting queens problem, *Amer. Math. Monthly* **76** (1969) 399–400; *MR* **39** #4018.
J. L. Selfridge, Pairings of the first $2n$ integers so that sums and differences are all distinct, *Notices Amer. Math. Soc.* **10** (1963) 195.
Mok-Kong Shen and Tsen-Pao Shen, Research Problem 39, *Bull. Amer. Math. Soc.* **68** (1962) 557.
M. Slater, Problem 1, *Bull. Amer. Math. Soc.* **69** (1963) 333.

**C19.** Selfridge calls a set of positive integers $a_1 < a_2 < \cdots < a_k$ **independent** if $\sum c_i a_i = 0$ (where the $c_i$ are integers, not all 0) implies that at least one of the $c_i$ is $< -1$. By using the pigeonhole principle it is easy to show that if $k$ positive integers are independent, then $a_1$ is at least $2^{k-1}$. He offers \$10.00 for an answer to the question: is the set of $k$ independent integers $a_i = 2^k - 2^{k-i}$ $(1 \le i \le k)$ the only set with largest member less than $2^k$? It is the only such set with $a_1 = 2^{k-1}$.

Call a(n infinite) sequence $\{a_i\}$ of positive integers **weakly independent** if any relation $\sum \varepsilon_i a_i = 0$ with $\varepsilon_i = 0$ or $\pm 1$ and $\varepsilon_i = 0$ except finitely often, implies $\varepsilon_i = 0$ for all $i$, and call it **strongly independent** if the same is true with $\varepsilon_i = 0, \pm 1$, or $\pm 2$. Richard Hall asks if every weakly independent sequence is a finite union of strongly independent sequences.

J. L. Selfridge, Problem 123, *Pi Mu Epsilon J.* **3** (1959–64) 118, 413–414.

**C20.** Paul Turán asks for a characterization of those positive integers $n$ which can be represented as the sum of four pairwise coprime squares, i.e., $n = x_1^2 + x_2^2 + x_3^2 + x_4^2$ with $(x_i, x_j) = 1$ $(1 \le i < j \le 4)$. If $8 | n$, $n$ cannot be so represented, and George Turán has proved that numbers $n \equiv 5 \pmod 6$ are not so representable.

On the other hand, Paul Turán conjectures that all positive integers can be represented as the sum of at most five pairwise coprime squares. Are all sufficiently large integers representable as the sum of *exactly* five pairwise coprime squares?

I. Chowla conjectures that every positive integer is the sum of at most four elements of the set $\{(p^2-1)/24 | p \text{ prime}, p \ge 5\}$. The smallest number which requires four such summands is 33.

Compare these problems with some earlier results of Wright, who showed, for example, that if $\lambda_1, \ldots, \lambda_4$ were given real numbers with sum 1, then every $n$ with a sufficiently large odd factor is expressible as $n = m_1^2 + \cdots + m_4^2$ with $|m_i^2 - \lambda_i n| = o(n)$. He has similar results for 5 or more squares, and for 3 squares (provided, of course, that $n$ is not of the form $4^a(8l+7)$ in this last case).

Bohman, Fröberg and Riesel showed that there are 31 numbers which *can't* be expressed as the sum of *distinct* squares, and that all numbers greater than 188 *can* be expressed as the sum of at most five distinct squares. Only two numbers, 124 and 188, require six distinct squares.

Jan Bohman, Carl-Erik Fröberg, and Hans Riesel, Partitions in squares, *BIT*, **19** (1979) 297–301; *MR* **80k**:10043.

E. M. Wright, The representation of a number as a sum of five or more squares, *Quart. J. Math.* (Oxford) **4** (1933) 37–51 and 228–232.

E. M. Wright, The representation of a number as a sum of four 'almost proportional' squares, *ibid.* **7** (1936) 230–240.

E. M. Wright, Representation of a number as a sum of three or four squares, *Proc. London Math. Soc.* (2) **42** (1937) 481–500.

# D. Some Diophantine Equations

> "A subject which can be described briefly by saying that a great part of it is concerned with the discussion of the rational or integer solutions of a polynomial equation $f(x_1, x_2, \ldots, x_n) = 0$ with integer coefficients. It is well-known that for many centuries, no other topic has engaged the attention of so many mathematicians, both professional and amateur, or has resulted in so many published papers."

This quotation from the preface of Mordell's book, *Diophantine Equations*, Academic Press, London, 1969, indicates that in this section we shall have to be even more eclectic than elsewhere. If you're interested in the subject, consult Mordell's book, which is a thoroughgoing but readable account of what is known, together with a great number of unsolved problems. There are well-developed theories of rational points on algebraic curves, so we mainly confine ourselves to higher dimensions, for which standard methods have not yet been developed.

## D1.

> "It has seemed to many Geometers that this theorem [Fermat's Last "Theorem"] may be generalized. Just as there do not exist two cubes whose sum or difference is a cube, it is certain that it is impossible to exhibit three biquadrates whose sum is a biquadrate, but that at least four biquadrates are needed if their sum is to be a biquadrate, although no one has been able up to the present to assign four such biquadrates. In the same manner it would seem to be impossible to exhibit four fifth powers whose sum is a fifth power, and similarly for higher powers."

No advance was made on Euler's statement until 1911 when R. Norrie assigned four such biquadrates:

$$30^4 + 120^4 + 272^4 + 315^4 = 353^4$$

Fifty-five years later Lander and Parkin gave a counter-example to Euler's more general conjecture:

$$27^5 + 84^5 + 110^5 + 133^5 = 144^5$$

It has not been proved that $a^4 + b^4 + c^4 = d^4$ has no solution in integers; there is none with $d \leq 220{,}000$. In fact even $a^4 + b^4 + c^4 = d^2$ is unknown.

Simcha Brudno asks the following questions. Is there a parametric solution to $a^5 + b^5 + c^5 + d^5 = e^5$? (There is but one solution with $e \leq 765$.) Is there a parametric solution to $a^4 + b^4 + c^4 + d^4 = e^4$? Are there counter-examples to Euler's conjecture with higher powers? Is there a solution of $a^6 + b^6 + c^6 + d^6 + e^6 = f^6$? Parametric solutions are known for equal sums of equal numbers of like powers,

$$\sum_{i=1}^{m} a_i^s = \sum_{i=1}^{m} b_i^s$$

with $a_i > 0$, $b_i > 0$, for $2 \leq s \leq 4$ and $m \leq 2$ and for $s = 5,6$ and $m \leq 3$. Can a solution be found for $s = 7$ and $m = 4$? For $s = 5$, $m = 2$, it is not known if there is any nontrivial solution of $a^5 + b^5 = c^5 + d^5$.

A method is known for generating parametric solutions of $a^4 + b^4 = c^4 + d^4$ which will generate all published solutions from the trivial one $(\lambda, 1, \lambda, 1)$; it will only produce solutions of degree $6n + 1$. Here, in answer to a question of Brudno, $6n + 1$ need not be prime. Although degree 25 does not appear, 49 does.

Swinnerton-Dyer has a second method for generating new solutions from old and can show that the two methods, together with the symmetries, generate all *nonsingular* parametric solutions, i.e., all solutions which correspond to nonsingular curves. Moreover, the process is constructive in the sense that he can give a finite procedure for finding all nonsingular solutions of given degree. All nonsingular solutions have odd degree, and all sufficiently large odd degrees do occur. Unfortunately, singular solutions do exist. Swinnerton-Dyer has a process for generating them, but has no reason to believe that it gives them all. The problem of describing them all needs completely new ideas. Some of the singular solutions have even degree and he conjectures (and could probably prove) that all sufficiently large even degrees occur in this way.

In the same sense, Andrew Bremner can find "all" parametric solutions of $a^6 + b^6 + c^6 = d^6 + e^6 + f^6$ which also satisfy the equations

$$a^2 + ad - d^2 = f^2 + fc - c^2$$
$$b^2 + be - e^2 = d^2 + da - a^2$$
$$c^2 + cf - f^2 = e^2 + eb - b^2$$

(this is not such a restriction as might at first appear). He can also find "all" parametric solutions of $a^5 + b^5 + c^5 = d^5 + e^5 + f^5$ which also satisfy $a + b + c = d + e + f$ and $a - b = d - e$.

B. J. Birch and H. P. F. Swinnerton-Dyer, Notes on elliptic curves, II, *J. reine angew. Math.* **218** (1965) 79–108.
Andrew Bremner, Pythagorean triangles and a quartic surface, *J. reine angew. Math.* **318** (1980) 120–125.
Andrew Bremner, A geometric approach to equal sums of sixth powers, *Proc. London Math. Soc.* (to appear).
Andrew Bremner, A geometric approach to equal sums of fifth powers, *J. Number Theory* (to appear).
S. Brudno, Some new results on equal sums of like powers, *Math. Comput.* **23** (1969) 877–880.
S. Brudno, On generating infinitely many solutions of the diophantine equation $A^6 + B^6 + C^6 = D^6 + E^6 + F^6$, *Math. Comput.* **24** (1970) 453–454.
S. Brudno, Problem 4, *Proc. Number Theory Conf. Univ. of Colorado*, Boulder, 1972, 256–257.
Simcha Brudno, Triples of sixth powers with equal sums, *Math. Comput.* **30** (1976) 646–648.
S. Brudno and I. Kaplansky, Equal sums of sixth powers, *J. Number Theory* **6** (1974) 401–403.
V. A. Dem'janenko, L. Euler's conjecture (Russian), *Acta Arith.* **25** (1973/74) 127–135; *MR* **50** #12912.
Jan Kubiček, A simple new solution to the diophantine equation $A^3 + B^3 + C^3 = D^3$, (Czech, German summary), *Časopis Pěst. Mat.* **99** (1974) 177–178.
L. J. Lander, Geometric aspects of diophantine equations involving equal sums of like powers, *Amer. Math Monthly* **75** (1968) 1061–1073.
L. J. Lander and T. R. Parkin, Counterexample to Euler's conjecture on sums of like powers, *Bull. Amer. Math. Soc.* **72** (1966) 1079; *MR* **33** #5554.
L. J. Lander, T. R. Parkin, and J. L. Selfridge, A survey of equal sums of like powers, *Math. Comput.* **21** (1967) 446–459; *MR* **36** #5060.
R. Norrie, *Univ. of St. Andrews 500th Anniv. Mem. Vol.*, Edinburgh, 1911, 89.
Morgan Ward, Euler's three biquadrate problem, *Proc. Nat. Acad. Sci. U.S.A.* **31** (1945) 125–127; *MR* **6**, 259.
Morgan Ward, Euler's problem on sums of three fourth powers, *Duke Math. J.* **15** (1948) 827–837; *MR* **10**, 283.

**D2.** J. M. Gandhi has tirelessly investigated the Fermat problem; here are a few of the many subsidiary problems which could possibly be more tractable than the original.

Are there integers $c$ for which $x^4 + y^4 = cz^4$ has integer solutions with $(x, y) = 1$?

Prove that $x^{11} + y^{11} = 6z^{11}$ has no integer solutions with none of $x, y, z$ divisible by 11.

Prove that $x^{11} + y^{11} = 3z^{11}$ and $x^{11} + y^{11} = 9z^{11}$ have no integer solutions with $11 | x$, $11 \nmid yz$.

Prove that $x^n + y^n = n!z^n$ has no integer solution. Erdös and Obláth showed that $x^p \pm y^p = n!$ has none.

Prove that $x^9 + y^9 = 7z^9$ is impossible in integers if $3 | z$.

Harold M. Edwards, *Fermat's Last Theorem, a Genetic Introduction to Algebraic Number Theory*, Springer-Verlag, New York, 1977.

P. Erdös and R. Obláth, Über diophantische Gleichungen der Form $n! = x^p \pm y^p$ and $n! \pm m! = x^p$, *Acta Litt. Sci. Szeged* **8** (1937) 241–255; *Zbl.* 17.004.
K. Inkeri and A. J. van der Poorten, Some remarks on Fermat's conjecture, *Acta Arith.* **36** (1980) 107–111.
Wells Johnson, Irregular primes and cyclotomic invariants, *Math. Comput.* **29** (1975) 113–120; *MR* **51** #12781.
D. H. Lehmer, On Fermat's quotient, base two, *Math. Comput.* **36** (1981) 289–290.
Paulo Ribenboim, *13 Lectures on Fermat's Last Theorem*, Springer-Verlag, New York, Heidelberg, Berlin, 1979; see *Bull. Amer. Math. Soc.* **4** (1981) 218–222; *MR* **81f**: 10023.
J. L. Selfridge, C. A. Nicol and H. S. Vandiver, Proof of Fermat's last theorem for all prime exponents less than 4002, *Proc. Nat. Acad. Sci. U.S.A.* **41** (1955) 970–973; *MR* **17**, 348.
Daniel Shanks and H. C. Williams, Gunderson's function in Fermat's last theorem, *Math. Comput.* **36** (1981) 291–295.
Samuel S. Wagstaff, The irregular primes to 125000, *Math. Comput.* **32** (1978) 583–591; *MR* **58** #10711.

**D3.** Mordell, on page 259 of his book, asks if the only integer solutions of

$$6y^2 = (x + 1)(x^2 - x + 6)$$

are given by $x = -1, 0, 2, 7, 15$, and 74? By Theorem 1 of Mordell's Chapter 27 there are only finitely many. The equation arises from

$$y^2 = \binom{x}{0} + \binom{x}{1} + \binom{x}{2} + \binom{x}{3}.$$

Similarly, Martin Gardner took the figurate numbers: triangle, square, tetrahedron and square pyramid; and equated them in pairs. Of the six resulting problems, he noted that they were all solved except "triangle = square pyramid," which leads to the equation

$$3(2y + 1)^2 = 8x^3 + 12x^2 + 4x + 3.$$

The number of solutions is again finite. Are they all given by $x = -1, 0, 1, 5, 6$, and 85?

The case "square pyramid = square" is Lucas's problem. Is $x = 24$, $y = 70$ the only nontrivial solution of the diophantine equation

$$y^2 = x(x + 1)(2x + 1)/6?$$

This was solved (affirmatively) by Watson, using elliptic functions, and by Ljunggren, using a Pell equation in a quadratic field. Mordell asked if there was an elementary proof.

The same equation in disguise is to ask if (48, 140) is the unique nontrivial solution to the case "square = tetrahedron," since the previous equation may be written

$$(2y)^2 = 2x(2x + 1)(2x + 2)/6.$$

H. E. Dudeney, *Amusements in Mathematics*, Nelson, 1917, 26, 167.
Raphael Finkelstein, On a diophantine equation with no non-trivial integral solution, *Amer. Math. Monthly* **73** (1966) 471–477.

W. Ljunggren, New solution of a problem proposed by E. Lucas, *Norsk Mat. Tidskr.* **34** (1952) 65–72.
E. Lucas, Problem 1180, *Nouv. Ann. Math.* (2) **14** (1875) 336.
G. N. Watson, The problem of the square pyramid, *Messenger of Math.* **48** (1918/19) 1–22.

**D4.** Let $r_{k,l}(n)$ be the number of solutions of $n = \sum_{i=1}^{l} x_i^k$ in *positive* integers $x_i$. Hardy and Littlewood's Hypothesis K is that $\varepsilon > 0$ implies that $r_{k,k}(n) = O(n^{\varepsilon})$. This is well-known for $k = 2$; in fact, for sufficiently large $n$,

$$r_{2,2}(n) < n^{(1+\varepsilon)\ln 2/\ln\ln n}$$

and this does not hold if ln 2 is replaced by anything smaller. Mahler disproved the hypothesis for $k = 3$ by showing that $r_{3,3}(n) > c_1 n^{1/12}$ for infinitely many $n$.

Erdös thinks it possible that for all $n$, $r_{3,3}(n) < c_2 n^{1/12}$ but nothing is known. Probably Hypothesis K fails for every $k \geq 3$, but also it's probable that $\sum_{n=1}^{x} (r_{k,k}(n))^2 < x^{1+\varepsilon}$ for sufficiently large $x$.

S. Chowla proved that for $k \geq 5$, $r_{k,k}(n) \neq O(1)$ and, with Erdös, that for every $k \geq 2$ and for infinitely many $n$,

$$r_{k,k}(n) > \exp(c_k \ln n/\ln\ln n).$$

Mordell proved $r_{3,2}(n) \neq O(1)$ and Mahler that $r_{3,2}(n) > (\ln n)^{1/4}$ for infinitely many $n$. No nontrivial upper bound for $r_{3,2}(n)$ is known. Jean Lagrange has shown that $\limsup r_{4,2}(n) \geq 2$ and $\limsup r_{4,3}(n) = \infty$.

Another tough problem is to estimate $A_{k,l}(x)$, the number of $n \leq x$ which are expressible as the sum of $l$ $k$th powers. Landau showed

$$A_{2,2}(x) = (c + o(1))x/(\ln x)^{1/2},$$

Erdös and Mahler proved that if $k > 2$, then $A_{k,2} > c_k x^{2/k}$, and Hooley that $A_{k,2} = (c_k + o(1))x^{2/k}$. It seems certain that if $l < k$, then $A_{k,l} > c_{k,l} x^{l/k}$ and that $A_{k,k} > x^{1-\varepsilon}$ for every $\varepsilon$, but these have not been established.

It follows from the Chowla–Erdös result that for all $k$ there is an $n_k$ such that the number of solutions of $n_k = p^3 + q^3 + r^3$ is greater than $k$. No corresponding result is known for more than three summands.

S. Chowla, The number of representations of a large number as a sum of non-negative $n$th powers, *Indian Phys.-Math. J.* **6** (1935) 65–68; *Zbl.* 12.339.
H. Davenport, Sums of three positive cubes, *J. London Math. Soc.* **25** (1950) 339–343; *MR* **12**, 393.
P. Erdös, On the representation of an integer as the sum of $k$ $k$th powers, *J. London Math. Soc.* **11** (1936) 133–136; *Zbl.* 13.390.
P. Erdös, On the sum and difference of squares of primes I, II, *J. London Math. Soc.* **12** (1937) 133–136, 168–171; *Zbl.* 16.201, 17.103.
P. Erdös and K. Mahler, On the number of integers which can be represented by a binary form, *J. London Math. Soc.* **13** (1938) 134–139.
P. Erdös and E. Szemerédi, On the number of solutions of $m = \sum_{i=1}^{k} x_i^k$, *Proc. Symp. Pure Math.* **24** Amer. Math. Soc., Providence, 1972, 83–90.

G. H. Hardy and J. E. Littlewood, Partitio Numerorum VI: Further researches in Waring's problem, *Math. Z.* **23** (1925) 1–37.
Jean Lagrange, Thèse d'Etat de l'Université de Reims, 1976.
K. Mahler, Note on hypothesis K of Hardy and Littlewoord, *J. London Math. Soc.* **11** (1936) 136–138.
K. Mahler, On the lattice points on curves of genus 1, *Proc. London Math. Soc.* **39** (1935) 431–466.

**D5.** Is every number the sum of four cubes? This has been proved for all numbers except those of the form $9n \pm 4$.

More demanding is to ask if every number is the sum of four cubes with two of them equal. Specifically, is there a solution of $76 = x^3 + y^3 + 2z^3$? The other numbers less than 1000 which are still in doubt are 148, 183, 230, 253, 356, 418, 428, 445, 482, 491, 519, 580, 671, 734, 788, 923, 931, and 967.

Are all numbers which are not of the form $9n \pm 4$ the sum of *three* cubes? In particular, does $30 = x^3 + y^3 + z^3$ have a solution?

The equation $3 = x^3 + y^3 + z^3$ has the solutions $(1, 1, 1)$ and $(4, 4, -5)$; how many others are there?

W. J. Ellison, Waring's problem, *Amer. Math. Monthly* **78** (1971) 10–36.
Chao Ko, Decompositions into four cubes, *J. London Math. Soc.* **11** (1936) 218–219.
V. L. Gardiner, R. B. Lazarus, and P. R. Stein, Solutions of the diophantine equation $x^3 + y^3 = z^3 - d$, *Math. Comput.* **18** (1964) 408–413; *MR* **31** #119.
M. Lal, W. Russell, and W. J. Blundon, A note on sums of four cubes, *Math. Comput.* **23** (1969) 423–424; *MR* **39** #6819.
A Mąkowski, Sur quelques problèmes concernant les sommes de quatre cubes, *Acta Arith.* **5** (1959) 121–123; *MR* **21** #5609.
J. C. P. Miller and M. F. C. Woollett, Solutions of the diophantine equation $x^3 + y^3 + z^3 = k$, *J. London Math. Soc.* **30** (1955) 101–110; *MR* **16**, 979.
A. Schinzel and W. Sierpiński, Sur les sommes de quatre cubes, *Acta Arith.* **4** (1958) 20–30.

**D6.** Ljunggren has shown that the only solutions of $x^2 = 2y^4 - 1$ in positive integers are $(1, 1)$ and $(239, 13)$ but his proof is difficult. Mordell asks if it is possible to find a simple or elementary proof.

Ljunggren and others have made considerable investigations into equations of similar type. A selection of references is given below. Cohn can now handle the equation $y^2 = Dx^4 + 1$ for all $D \le 400$.

Edward A. Bender and Norman P. Herzberg, Some diophantine equations related to the quadratic form $ax^2 + by^2$, *Bull. Amer. Math. Soc.* **81** (1975) 161–162.
J. Blass, On the diophantine equation $Y^2 + K = X^5$, *Bull. Amer. Math. Soc.* **80** (1974) 329.
J. H. E. Cohn, On square Fibonacci numbers, *J. London Math. Soc.* **39** (1964) 537–540; *MR* **29** #1166.
J. H. E. Cohn, The diophantine equation $y^2 = Dx^4 + 1$, I, *J. London Math. Soc.* **42** (1967) 475–476; *MR* **35** #4158; II, *Acta Arith.* **28** (1975/76) 273–275; *MR* **52** #8029; III, *Math. Scand.* **42** (1978) 180–188; *MR* **80a**:10031.

N. P. Herzberg, Integer solutions of $by^2 + p^n = x^3$, *J. Number Theory* **7** (1975) 221–234; *Zbl.* 302.10021.
D. J. Lewis, Two classes of diophantine equations, *Pacific J. Math.* **11** (1961) 1063–1076.
W. Ljunggren, Zur Theorie der Gleichung $x^2 + 1 = Dy^4$, *Avh. Norske Vid. Akad. Oslo*, I, **5** (1942) #5, 27 pp; *MR* **8**, 6.
W. Ljunggren, On a diophantine equation, *Norske Vid. Selsk. Forh. Trondheim*, **18** #32 (1945) 125–128; *MR* **8**, 136.
W. Ljunggren, New theorems concerning the diophantine equation $Cx^2 + D = y^n$, *Norske Vid. Selsk. Forh. Trondheim* **29** (1956) 1–4; *MR* **17**, 1185.
W. Ljunggren, On the diophantine equation $Cx^2 + D = y^n$, *Pacific J. Math.* **14** (1964) 585–596; *MR* **28** #5035.
W. Ljunggren, Some remarks on the diophantine equation $x^2 - Dy^4 = 1$ and $x^4 - Dy^2 = 1$, *J. London Math. Soc.* **41** (1966) 542–544; *MR* **33** #5555.
L. J. Mordell, The diophantine equation $y^2 = Dx^4 + 1$, *J. London Math. Soc.* **39** (1964) 161–164; *MR* **29** #65.
T. Nagell, Contributions to the theory of a category of diophantine equations of the second degree with two unknowns, *Nova Acta Soc. Sci. Upsal.* (4) **16** (1955) #2; 38 pp; *MR* **17**, 13.

**D7.** Leo Moser investigated the equation

$$1^n + 2^n + \cdots + (m-1)^n = m^n$$

which the late Rufus Bowen conjectured to have no nontrivial solutions, and showed that there were none with $m \le 10^{10^6}$. Of course $n \sim m \ln 2$. Robert Tijdeman observes that general results on the equation

$$1^n + 2^n + \cdots + k^n = m^n$$

have no implications for the special equation, and gives the following recent history.

Van de Lune proved that

$$\text{(L)} \qquad 1^n + 2^n + \cdots + (m-1)^n < m^n$$

if $(m-1)^n < \frac{1}{2}m^n$. Best and te Riele proved that

$$\text{(M)} \qquad 1^n + 2^n + \cdots + (m-1)^n > m^n$$

if $(m-2)^n \ge \frac{1}{2}(m-1)^n$. We may therefore assume that for any $m$ the integer $n$ is determined by

$$\left(1 - \frac{1}{m}\right)^n > \frac{1}{2} > \left(1 - \frac{1}{m-1}\right)^n.$$

Erdös conjectured that both (L) and (M) hold infinitely often. Van de Lune and te Riele proved that (M) holds for almost all $m$ and Best and te Riele proved that (L) holds for at most $\ln x$ values of $m \le x$. They computed 33 pairs $(m, n)$ for which (L) holds, the smallest being

$$m = 1121\ 626023\ 352385, \qquad n = 777\ 451915\ 729368$$

It seems to be hard to show that (L) holds infinitely often.

M. R. Best and H. J. J. te Riele, On a conjecture of Erdös concerning sums of powers of integers, *Report NW 23/76*, Mathematisch Centrum Amsterdam, 1976.
P. Erdös, Advanced problem 4347, *Amer. Math. Monthly* **56** (1949) 343.
K. Györy, R. Tijdeman and M. Voorhoeve, On the equation $1^k + 2^k + \cdots + x^k = y^z$, *Acta Arith.* **37** (1980) 233–240.
J. van de Lune, On a conjecture of Erdös (I), *Report ZW 54/75*, Mathematisch Centrum, Amsterdam, 1975.
J. van de Lune and H. J. J. te Riele, On a conjecture of Erdös (II), *Report ZW 56/75*, Mathematisch Centrum, Amsterdam, 1975.
L. Moser, On the diophantine Equation $1^n + 2^n + \cdots + (m-1)^n = m^n$, *Scripta Math.* **19** (1953) 84–88; *MR* **14**–950.
J. J. Schäffer, The equation $1^p + 2^p + \cdots + n^p = m^q$, *Acta Math.* **95** (1956) 155–189; *MR* **17**, 1187.
M. Voorhoeve, K. Györy and R. Tijdeman, On the diophantine equation $1^k + 2^k + \cdots + x^k + R(x) = y^z$, *Acta Math.* **143** (1979) 1–8; *MR* **80e**:10020.

**D8.** Wunderlich asks for (a parametric representation of) *all* solutions of the equation $x^3 + y^3 + z^3 = x + y + z$. Bernstein, S. Chowla, Edgar, Fraenkel, Oppenheim, Segal, and Sierpiński have given solutions, some of them parametric, so there are certainly infinitely many. Eighty-eight of them have unknowns less than 13000. A general solution has not been given.

Leon Bernstein, Explicit solutions of pyramidal Diophantine equations, *Canad. Math. Bull.* **15** (1972) 177–184; *MR* **46** #3442.
Hugh Maxwell Edgar, Some remarks on the Diophantine equation $x^3 + y^3 + z^3 = x + y + z$, *Proc. Amer. Math. Soc.* **16** (1965) 148–153; *MR* **30** #1094.
A. S. Fraenkel, Diophantine equations involving generalized triangular and tetrahedral numbers, in *Computers in Number Theory, Proc. Atlas Symp. No. 2*, Oxford 1969 Academic Press, London and New York (1971) 99–114.
A. Oppenheim, On the Diophantine equation $x^3 + y^3 + z^3 = x + y + z$, *Proc. Amer. Math. Soc.* **17** (1966) 493–496; *MR* **32** #5590.
A. Oppenheim, On the diophantine equation $x^3 + y^3 - z^3 = px + py - qz$, *Univ. Beograd Publ. Elektrotehn. Fak. Ser.* #235 (1968); *MR* **39** #126.
S. L. Segal, A note on pyramidal numbers, *Amer. Math. Monthly* **69** (1962) 637–638; *Zbl.* **105**, 36.
W. Sierpiński, Sur une propriété des nombres tétraédraux, *Elem. Math.* **17** (1962) 29–30; *MR* **24** #A 3118.
W. Sierpiński, Trois nombres tétraédraux en progression arithmetique, *Elem. Math.* **18** (1963) 54–55; *MR* **26** #4957.
M. Wunderlich, Certain properties of pyramidal and figurate numbers, *Math. Comput.* **16** (1962) 482–486; *MR* **26** #6115.

**D9.** Except that there remains a finite amount of computation, Tijdeman has settled the old conjecture of Catalan, that the only consecutive powers higher than the first are $2^3$ and $3^2$. In fact, if $a_1 = 4$, $a_2 = 8$, $a_3 = 9, \ldots$ is the sequence of such powers, Choodnowski claims to have proved that $a_{n+1} - a_n$ tends to infinity with $n$. Erdös conjectures that $a_{n+1} - a_n > c'n^c$, but says there is no present hope of proof.

Jan Mycielski observes that all numbers other than those congruent to 2 (mod 4) can be expressed as the difference of two powers each greater than the first. He asks if 6, 14, or 34 can be so expressed.

Erdös asks if there are infinitely many numbers not of the form $x^k - y^l$, $k > 1, l > 1$.

Carl Rudnick denotes by $N(r)$ the number of positive solutions of $x^4 - y^4 = r$, and asks if $N(r)$ is bounded. H. Hansraj Gupta observes that Hardy and Wright (p. 201) give a parametric solution of $x^4 - y^4 = u^4 - v^4$, which establishes that $N(r)$ is 0, 1, or 2 infinitely often. For example $133^4 - 59^4 = 158^4 - 134^4 = 300783360$. If $N(r) = 3$, then $r$ must be very, very large. But there can hardly be any doubt that $N(r)$ is bounded.

Hugh Edgar asks how many solutions $(m, n)$ does $p^m - q^n = 2^h$ have, for primes $p$, $q$ and integer $h$. At most one? Only finitely many? Examples are $3^2 - 2^3 = 2^0$; $5^3 - 11^2 = 2^2$; $5^2 - 3^2 = 2^4$.

R. Tijdeman, On the equation of Catalan, *Acta Arith.* **29** (1976) 197–209; *MR* **53** #7941.

**D10.** Brenner and Foster pose the following general problem. Let $\{p_i\}$ be a finite set of primes and $\varepsilon_i = \pm 1$. When can the exponential diophantine equation $\sum \varepsilon_i p_i^{x_i} = 0$ be solved by elementary methods (i.e. by modular arithmetic)? More exactly, given $p_i$, $\varepsilon_i$, what criteria determine whether there exists a modulus $M$ such that the given equation is equivalent to the congruence $\sum \varepsilon_i p_i^{x_i} \equiv 0 \pmod{M}$? They solve many particular cases, mostly where the $p_i$ are four in number and less than 108. In a few cases elementary methods avail, even if two of the primes are equal, but in general they do not. In fact, neither $3^a = 1 + 2^b + 2^c$ nor $2^a + 3^b = 2^c + 3^d$ can be reduced to a single congruence.

Hugh Edgar asks if there is a solution, other than $1 + 3 + 3^2 + 3^3 + 3^4 = 11^2$, of the equation $1 + q + q^2 + \cdots + q^{x-1} = p^y$ with $p$, $q$ odd primes and $x \geq 5$, $y \geq 2$.

Leo J. Alex, Problem E 2880, *Amer. Math. Monthly*, **88** (1981) 291.
J. L. Brenner and Lorraine L. Foster, Exponential diophantine equations, *Pacific J. Math.* (to appear).

**D11. Egyptian fractions.** The Rhind papyrus is amongst the oldest written mathematics which has come down to us; it concerns the representation of rational numbers as the sum of unit fractions,

$$\frac{m}{n} = \frac{1}{x_1} + \frac{1}{x_2} + \cdots + \frac{1}{x_k}.$$

This has suggested numerous problems, many of which are unsolved, and continues to suggest new problems, so that interest in Egyptian fractions is as great as it has ever been. We give an extensive bibliography, but this may only be a fraction of what has been written. Bleicher has given a careful survey of the subject and draws attention to the various algorithms that have been proposed for constructing representations of the given type: the Fibonacci–Sylvester algorithm, Erdös's algorithm, Golomb's algorithm and two of his own, the Farey series algorithm and the continued fraction algorithm. See also the extensive Section 4 of the collection of problems by Erdös and Graham mentioned at the beginning of this volume, and the bibliography obtainable from Paul Campbell.

Erdös and Straus conjectured that the equation

$$\frac{4}{n}=\frac{1}{x}+\frac{1}{y}+\frac{1}{z}$$

could be solved in positive integers for all $n>1$. There is a good account of the problem in Mordell's book, where it is shown that the conjecture is true, except possibly in cases where $n$ is prime and congruent to $1^2$, $11^2$, $13^2$, $17^2$, $19^2$, or $23^2$ (mod 840). Several have worked on the problem, including Bernstein, Obláth, Rosati, Shapiro, Straus, Yamamoto, and Nicola Franceschine who has verified the conjecture for $n \le 10^8$. Schinzel has observed that one can express

$$\frac{4}{at+b}=\frac{1}{x(t)}+\frac{1}{y(t)}+\frac{1}{z(t)}$$

with $x(t)$, $y(t)$, $z(t)$ integer polynomials in $t$, only if $b$ is *not* a quadratic residue of $a$.

Sierpiński made the corresponding conjecture concerning

$$\frac{5}{n}=\frac{1}{x}+\frac{1}{y}+\frac{1}{z}.$$

Palamà confirmed it for all $n \le 922321$ and Stewart has extended this to $n \le 1057438801$ and for all $n$ not of the form $278460k+1$.

Schinzel relaxed the condition that the integers $x$, $y$, $z$ should be positive, replaced the numerators 4 and 5 by a general $m$ and required the truth only for $n > n_m$. That $n_m$ may be greater than $m$ is exemplified by $n_{18}=23$. The conjecture has been established for successively larger values of $m$ by Schinzel, Sierpiński, Sedláček, Palamà and Stewart and Webb, who prove it for $m<36$. Breusch and Stewart independently showed that if $m/n>0$ and $n$ is odd, then $m/n$ is the sum of a finite number of reciprocals of odd integers. See also Graham's papers. Vaughan has shown that if $E_m(N)$ is the number of $n \le N$ for which $m/n = 1/x+1/y+1/z$ has no solution, then

$$E_m(N) \ll N \exp\{-c(\ln N)^{2/3}\}$$

where $c$ depends only on $m$.

In contrast to the result of Breusch and Stewart, the following problem, asked by Stein, Selfridge, Graham, and others, has not been solved. If $m/n$, a rational number ($n$ odd), is expressed as $\sum 1/x_i$, where the $x_i$ are successively chosen to be the least possible odd integers which leave a nonnegative remainder, is the sum always finite? For example,

$$\frac{2}{7}=\frac{1}{5}+\frac{1}{13}+\frac{1}{115}+\frac{1}{10465}$$

John Leech, in a letter dated 77:03:14, asks what is known about sets of unequal odd integers whose reciprocals add to 1, such as

$$\frac{1}{3}+\frac{1}{5}+\frac{1}{7}+\frac{1}{9}+\frac{1}{15}+\frac{1}{21}+\frac{1}{27}+\frac{1}{35}+\frac{1}{63}+\frac{1}{105}+\frac{1}{135}=1$$

He says that you need at least nine in the set, while on the other hand the largest denominator must be at least 105. Notice the connexion with Sierpiński's pseudoperfect numbers (B2)

$$945 = 315 + 189 + 135 + 105 + 63 + 45 + 35 + 27 + 15 + 9 + 7$$

It is known that $m/n$, $n$ odd, is always expressible as a sum of distinct odd unit fractions.

Erdös, in a letter dated 72:01:14, sets $\frac{1}{2}+\frac{1}{3}+\cdots+1/n = a/b$, where $b = [2, 3, \ldots, n]$, the l.c.m. of $2, 3, \ldots, n$. He observes that $\frac{1}{2}+\frac{1}{3}=\frac{5}{6}$ and that $\frac{1}{2}+\frac{1}{3}+\frac{1}{4}=\frac{13}{12}$ are such that $a \pm 1 \equiv 0 \pmod b$ and asks if this occurs again; he conjectures not. Is $(a, b) = 1$ infinitely often?

If $\sum_{i=1}^{t} 1/x_i = 1$ with $x_1 < x_2 < x_3 < \cdots$ *distinct* positive integers, Erdös and Graham ask what is $m(t)$, the min max $x_i$ where the minimum is taken over all sets $\{x_i\}$. For example, $m(3) = 6$, $m(4) = 12$, $m(12) = 120$. Is $m(t) < ct$ for some constant $c$?

In the notation of the last paragraph, is it possible to have $x_{i+1} - x_i \le 2$ for all $i$? Erdös conjectures that it is not and offers \$10.00 for a solution.

Given a sequence $x_1, x_2, \ldots$ of positive density, is there always a finite subset with $\sum 1/x_{i_k} = 1$? If $x_i < ci$ for all $i$, is there such a finite subset? Erdös again offers \$10.00 for a solution. If $\liminf x_i/i < \infty$, he strongly conjectures that the answer is negative and offers only \$5.00 for a solution.

Denote by $N(t)$ the number of solutions $\{x_1, x_2, \ldots, x_t\}$ of $1 = \sum 1/x_i$ and by $M(t)$ the number of distinct solutions, $x_1 \le x_2 \le \cdots \le x_t$. Singmaster calculated

| $t$ | 1 | 2 | 3 | 4 | 5 | 6 |
|---|---|---|---|---|---|---|
| $M(t)$ | 1 | 1 | 3 | 14 | 147 | 3462 |
| $N(t)$ | 1 | 1 | 10 | 215 | 12231 | 2025462 |

and Erdös asked for an asymptotic formula for $M(t)$ or $N(t)$.

Graham has shown that if $n > 77$ we can partition $n = x_1 + x_2 + \cdots + x_t$ into $t$ distinct positive integers so that $\sum 1/x_i = 1$. More generally, that for any positive rational numbers $\alpha$, $\beta$, there is an integer $r(\alpha, \beta)$, which we will take to be the least, such that any integer greater than $r$ can be partitioned into distinct positive integers greater than $\beta$, whose reciprocals sum to $\alpha$. Little is known about $r(\alpha, \beta)$, except that unpublished work of D. H. Lehmer shows that 77 cannot be partitioned in this way, so that $r(1, 1) = 77$.

Graham conjectures that for $n$ sufficiently large (about $10^4$?) we can similarly partition $n = x_1^2 + x_2^2 + \cdots + x_t^2$ with $\sum 1/x_i = 1$. We can also ask for a decomposition $n = p(x_1) + p(x_2) + \cdots + p(x_t)$ where $p(x)$ is a "reasonable" polynomial; for example, $x^2 + x$ is unreasonable since it takes only even values.

L.-S. Hahn asks: if the positive integers are partitioned into a finite number of sets in any way, is there always a set such that *any* positive rational number can be expressed as the sum of the reciprocals of a finite number of distinct members of it? Here it must always be possible to choose the set, independent of the rational number. If this is not possible, then given any rational number, can one always choose a set with this property? Now the set can depend on the rational number.

Erdös, in a letter dated 76:02:10, puts

$$1 = \frac{1}{x_1} + \frac{1}{x_2} + \cdots + \frac{1}{x_k} \quad \text{where } x_1 < x_2 < \cdots < x_k$$

and asks, if $k$ is fixed, what is the max $x_1$? If $k$ varies, which integers can be equal to $x_k$, the largest denominator? Not primes, and not several other integers; do the excluded integers have positive density? Density 1 even? Which integers can be $x_k$ or $x_{k-1}$? $x_k$ or $x_{k-1}$ or $x_{k-2}$? Is max $x_{i+1} - x_i > 2$? Does $x_k/x_1 \to \infty$ with $k$? If the $x_i$ are the union of $r$ blocks of consecutive integers, the number of solutions is finite and depends only on $r$. What is max $x_k$ as a function of $r$? (Perhaps $\frac{1}{2} + \frac{1}{3} + \frac{1}{7} + \frac{1}{43} + \cdots$ gives the maximum?).

Nagell showed that the sum of the reciprocals of an arithmetic progression is never an integer. See also the paper of Erdös and Niven.

A. V. Aho and N. J. A. Sloane, Some doubly exponential sequences, *Fibonacci Quart.* **11** (1973) 429–438; *MR* **49** #209.

A. Aigner, Brüche als Summe von Stammbrüchen, *J. reine angew. Math.* **214/215** (1964) 174–179.

P. J. van Albada and J. H. van Lint, Reciprocal bases for the integers, *Amer. Math. Monthly* **70** (1963) 170–174.

E. J. Barbeau, Computer Challenge corner: Problem 477: A brute force program, *J. Recreational Math.* **9** (1976/77) 30.

E. J. Barbeau, Expressing one as a sum of distinct reciprocals: comments and a bibliography, *Eureka* (Ottawa) **3** (1977) 178–181.

Leon Bernstein, Zur Lösung der diophantischen Gleichung $m/n = 1/x + 1/y + 1/z$ insbesondere im Falle $m = 4$, *J. reine angew. Math.* **211** (1962) 1–10; *MR* **26** #77.

M. N. Bleicher, A new algorithm for the expansion of Egyptian fractions, *J. Number Theory* **4** (1972) 342–382; *MR* **48** #2052.

M. N. Bleicher and P. Erdös, The number of distinct subsums of $\sum_1^N 1/i$, *Math. Comput.* **29** (1975) 29–42 (and see *Notices Amer. Math. Soc.* **20** (1973) A-516).

M. N. Bleicher and P. Erdös, Denominators of Egyptian fractions, *J. Number Theory* **8** (1976) 157–168; *MR* **53** #7925; II, *Illinois J. Math.* **20** (1976) 598–613; *MR* **54** #7359.

Robert Breusch, A special case of Egyptian fractions, Solution to Advanced Problem 4512, *Amer. Math. Monthly* **61** (1954) 200–201.

W. S. Burnside, *Theory of Groups of Finite Order*, 2nd ed. Cambridge University Press, London, 1911, reprinted Dover, New York, 1955, Note A, 461–462.

N. Burshtein, On distinct unit fractions whose sum equals 1, *Discrete Math.* **5** (1973) 201–206.

Paul J. Campbell, Bibliography of algorithms for Egyptian fractions (preprint) Beloit Coll. Beloit WI 53511, U.S.A.

J. W. S. Cassels, On the representation of integers as the sum of distinct summands taken from a fixed set, *Acta Sci. Math. Szeged* **21** (1960) 111–124.

A. B. Chace, *The Rhind Mathematical Papyrus*, M.A.A., Oberlin, 1927.
Robert Cohen, Egyptian fraction expansions, *Math. Mag.* **46** (1973) 76–80; *MR* **47** #3300.
D. Culpin and D. Griffiths, Egyptian fractions, *Math. Gaz.* **63** (1979) 49–51; *MR* **80d**: 10014.
D. R. Curtiss, On Kellogg's Diophantine problem, *Amer. Math. Monthly*, **29** (1922) 380–387.
L. E. Dickson, *History of the Theory of Numbers*, Vol. 2 Diophantine Analysis, Chelsea, New York, 1952, 688–691.
P. Erdös, Egy Kürschák-féle elemi számelméleti tétel áltadanosítása, *Mat. es Phys. Lapok* 39 (1932).
P. Erdös, On arithmetical properties of Lambert series, *J. Indian Math. Soc.* **12** (1948) 63–66.
P. Erdös, On a diophantine equation (Hungarian. Russian and English summaries), *Mat. Lapok* **1** (1950) 192–210; *MR* **13**, 208.
P. Erdös, On the irrationality of certain series, *Nederl. Akad. Wetensch.* (*Indag. Math.*) **60** (1957) 212–219.
P. Erdös, Sur certaines séries à valeur irrationnelle, *Enseignement Math.* **4** (1958) 93–100.
P. Erdös, *Quelques Problèmes de la Théorie des Nombres*, Monographie de l'Enseignement Math. No. 6, Geneva, 1963, problems 72–74.
P. Erdös, Comment on problem E2427, *Amer. Math. Monthly*, **81** (1974) 780–782.
P. Erdös, Some problems and results on the irrationality of the sum of infinite series, *J. Math. Sci.* **10** (1975) 1–7.
Paul Erdös and Ivan Niven, Some properties of partial sums of the harmonic series, *Bull. Amer. Math. Soc.* **52** (1946) 248–251; *MR* **7**, 413.
P. Erdös and S. Stein, Sums of distinct unit fractions, *Proc. Amer. Math. Soc.* **14** (1963) 126–131.
P. Erdös and E. G. Straus, On the irrationality of certain Ahmes series, *J. Indian Math. Soc.* **27** (1968) 129–133.
P. Erdös and E. G. Straus, Some number theoretic results, *Pacific J. Math.* **36** (1971) 635–646.
P. Erdös and E. G. Straus, Solution of problem E2232, *Amer. Math. Monthly* **78** (1971) 302–303.
P. Erdös and E. G. Straus, On the irrationality of certain series, *Pacific J. Math.* **55** (1974) 85–92; *MR* **51** #3069.
P. Erdös and E. G. Straus, Solution to problem 387, *Nieuw Arch. Wisk.* **23** (1975) 183.
Nicola Franceschine, Egyptian Fractions, MA Dissertation, Sonoma State Coll. CA, 1978.
S. W. Golomb, An algebraic algorithm for the representation problems of the Ahmes papyrus, *Amer. Math. Monthly* **69** (1962) 785–786.
S. W. Golomb, On the sums of the reciprocals of the Fermat numbers and related irrationalities, *Canad. J. Math.* **15** (1963) 475–478.
R. L. Graham, A theorem on partitions, *J. Austral. Math. Soc.* **4** (1963) 435–441.
R. L. Graham, On finite sums of unit fractions, *Proc. London Math. Soc.* (3) **14** (1964) 193–207; *MR* **28** #3968.
R. L. Graham, On finite sums of reciprocals of distinct $n$th powers, *Pacific J. Math.* **14** (1964) 85–92; *MR* **28** #3004.
L.-S. Hahn, Problem E2689, *Amer. Math. Monthly* **85** (1978) 47.
J. W. Hille, Decomposing fractions, *Math. Gaz.* **62** (1978) 51–52.
Ludwig Holzer, *Zahlentheorie* Teil III, Ausgewählte Kapitel der Zahlentheorie, Math.-Nat. Bibl. No. 14a, B. G. Teubner-Verlag, Leipzig, 1965, Sect. A, 1–27; *MR* **34** #4186.
Dag Magne Johannessen, On unit fractions II, *Nordisk mat. Tidskr.* **25–26** (1978) 85–90; *MR* **80a**:10010; *Zbl.* 384.10004.

Dag Magne Johannessen and T. V. Søhus, On unit fractions I, *ibid* **22** (1974) 103–107; *MR* **55** #252; *Zbl*. 291.10010.

Ralph W. Jollensten, A note on the Egyptian problem, *Congressus Numerantium XVII*, Proc. 7th S. E. Conf. Combin. Graph Theory, Comput. 1976, 351–364; *MR* **55** #2746.

O. D. Kellogg, On a diophantine problem, *Amer. Math. Monthly*, **28** (1921) 300–303.

E. Kiss, Quelques remarques sur une équation diophantienne (Romanian. French summary) *Acad. R. P. Romîne Fil. Cluj, Stud. Cerc. Mat.* **10** (1959) 59–62.

E. Kiss, Remarques relatives à la représentation des fractions subunitaires en somme des fractions ayant le numerateur égal à l'unité (Romanian) *Acad. R. P. Romîne Fil. Cluj, Stud. Cerc. Mat.* **11** (1960) 319–323.

Ladis D. Kovach, Ancient algorithms adapted to modern computers, *Math. Mag.* **37** (1964) 159–165.

József Kürschák, A harmonikus sorról, *Mat. es. Phys. Lapok* **27** (1918) 299–300.

Denis Lawson, Ancient Egypt revisited, *Math. Gaz.* **54** (1970) 293–296; *MR* **58** #10697.

P. Montgomery, Solution to Problem E2689, *Amer. Math. Monthly* **86** (1979) 224.

L. J. Mordell, *Diophantine Equations*, Academic Press, London, 1969, 287–290.

T. Nagell, *Skr. Norske Vid. Akad. Kristiania I*, 1923, no. 13 (1924) 10–15.

M. Nakayama, On the decomposition of a rational number into "Stammbrüche," *Tôhoku Math. J.* **46** (1939) 1–21.

James R. Newman, The Rhind Papyrus, in *The World of Mathematics*, Allen and Unwin, London, 1960, 169–178.

R. Obláth, Sur l'équation diophantienne $4/n = 1/x_1 + 1/x_2 + 1/x_3$, *Mathesis* **59** (1950) 308–316; *MR* **12**, 481.

J. C. Owings, Another proof of the Egyptian fraction theorem, *Amer. Math. Monthly* **75** (1968) 777–778.

G. Palamà, Su di una congettura di Sierpiński relativa alla possibilità in numeri naturali della $5/n = 1/x_1 + 1/x_2 + 1/x_3$, *Boll. Un. Mat. Ital.* (3) **13** (1958) 65–72; *MR* **20** #3821.

G. Palamà, Su di una congettura di Schinzel, *Boll. Un. Mat. Ital.* (3) **14** (1959) 82–94; *MR* **22** #7989.

T. E. Peet, *The Rhind Mathematical Papyrus*, Univ. Press of Liverpool, London, 1923.

L. Pisano, *Scritti*, Vol. 1, B. Boncompagni, Rome, 1857.

Y. Rav, On the representation of a rational number as a sum of a fixed number of unit fractions, *J. reine angew. Math.* **222** (1966) 207–213.

L. A. Rosati, Sull'equazione diofantea $4/n = 1/x_1 + 1/x_2 + 1/x_3$, *Boll. Un. Mat. Ital.* (3) **9** (1954) 59–63; *MR* **15**, 684.

H. D. Ruderman, Problem E2232, *Amer. Math. Monthly* **77** (1970) 403.

Harry Ruderman, Bounds for Egyptian fraction partitions of unity, Problem E2427, *Amer. Math. Monthly* **80** (1973) 807.

H. E. Salzer, The approximation of numbers as sums of reciprocals, *Amer. Math. Monthly* **54** (1947) 135–142; *MR* **8**, 534.

H. E. Salzer, Further remarks on the approximation of numbers as sums of reciprocals, *Amer. Math. Monthly* **55** (1948) 350–356; *MR* **10**, 18.

Andrzej Schinzel, Sur quelques propriétés des nombres $3/n$ et $4/n$, où $n$ est un nombre impair, *Mathesis* **65** (1956) 219–222; *MR* **18**, 284.

Jiří Sedláček, Über die Stammbrüche, *Časopis Pěst. Mat.* **84** (1959) 188–197; *MR* **23** #A829.

Ernest S. Selmer, Unit fraction expansions and a multiplicative analog, *Nordisk mat. Tidskr.* **25–26** (1978) 91–109; *Zbl*. 384.10005.

W. Sierpiński, Sur les décompositions de nombres rationnels en fractions primaires, *Mathesis* **65** (1956) 16–32; *MR* **17**, 1185.

W. Sierpiński, *On the Decomposition of Rational Numbers into Unit Fractions* (Polish), Pánstwowe Wydawnictwo Naukowe, Warsaw, 1957.

W. Sierpiński, Sur une algorithme pour le développer les nombres réels en séries rapidement convergentes, *Bull. Int. Acad. Sci. Cracovie Ser. A Sci. Mat.* **8** (1911) 113–117.
David Singmaster, The number of representations of one as a sum of unit fractions (mimeographed note) 1972.
N. J. A. Sloane, *A Handbook of Integer Sequences*, Academic Press, New York, 1973.
B. M. Stewart, Sums of distinct divisors, *Amer. J. Math.* **76** (1954) 779–785; *MR* **16**, 336.
B. M. Stewart, *Theory of Numbers*, Macmillan, N.Y., 1964, 198–207.
B. M. Stewart and W. A. Webb, Sums of fractions with bounded numerators, *Canad. J. Math.* **18** (1966) 999–1003; *MR* **33** #7297.
E. G. Straus and M. V. Subbarao, On the representation of fractions as sum and difference of three simple fractions, *Congressus Numeratium XX*, Proc. 7th Conf. Numerical Math. Comput. Manitoba 1977, 561–579.
J. J. Sylvester, On a point in the theory of vulgar fractions, *Amer. J. Math.* **3** (1880) 332–335, 388–389.
D. G. Terzi, On a conjecture of Erdös-Straus, *BIT* **11** (1971) 212–216.
L. Theisinger, Bemerkung über die harmonische Reihe, *Monat. für Math. u. Physik* **26** (1915) 132–134.
R. C. Vaughan, On a problem of Erdös, Straus and Schinzel, *Mathematika* **17** (1970) 193–198.
C. Viola, On the diophantine equations $\prod_0^k x_i - \sum_0^k x_i = n$ and $\sum_0^k 1/x_i = a/n$, *Acta Arith.* **22** (1972/73) 339–352.
W. A. Webb, On $4/n = 1/x + 1/y + 1/z$, *Proc. Amer. Math. Soc.* **25** (1970) 578–584.
William A. Webb, Rationals not expressible as a sum of three unit fractions, *Elem. Math.* **29** (1974) 1–6.
William A. Webb, On a theorem of Rav concerning Egyptian fractions, *Canad. Math. Bull.* **18** (1975) 155–156.
William A. Webb, On the unsolvability of $k/n = 1/x + 1/y + 1/z$, *Notices Amer. Math. Soc.* **22** (1975) A-485.
William A. Webb, On the diophantine equation $k/n = a_1/x_1 + a_2/x_2 + a_3/x_3$ (loose Russian summary), *Časopis Pěst. Mat.* **101** (1976) 360–365.
H. S. Wilf, Reciprocal bases for the integers, Res. Problem 6, *Bull. Amer. Math. Soc.* **67** (1961) 456.
R. T. Worley, Signed sums of reciprocals I, II, *J. Australian Math. Soc.* **21** (1976) 410–414, 415–417.
Koichi Yamamoto, On a conjecture of Erdös, *Mem. Fac. Sci. Kyushū Univ.* Ser. A, **18** (1964) 166–167; *MR* **30** #1968.
K. Yamamoto, On the diophantine equation $4/n = 1/x + 1/y + 1/z$, *Mem. Fac. Sci. Kyushū Univ.* Ser. A, **19** (1965) 37–47.

**D12.** A diophantine equation which has excited a great deal of interest is

$$x^2 + y^2 + z^2 = 3xyz.$$

It obviously has the **singular solutions** (1, 1, 1) and (2, 1, 1), and all solutions can be generated from these since the equation is a quadratic in each of the variables, so one integer solution leads to a second, and it can be shown that, apart from the singular solutions, all solutions have distinct values of $x$, $y$, and $z$, so that each such solution is a **neighbor** of just three others (Figure 10). The numbers 1, 2, 5, 13, 29, 34, 89, 169, 194, 233, 433, 610, 985, . . . are called **Markoff numbers**. An outstanding problem is whether Figure 10 is a genuine

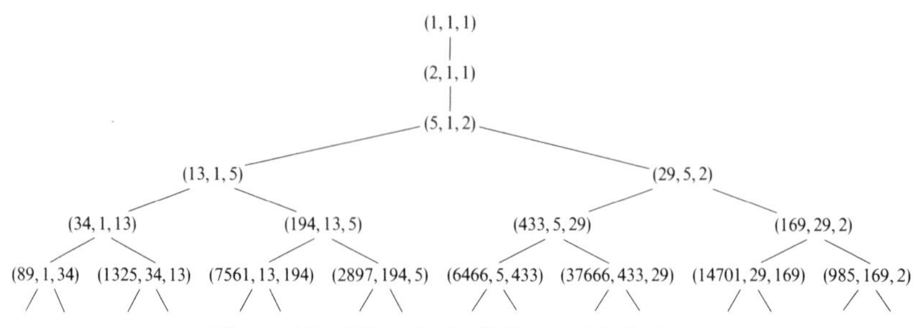


Figure 10. The Markoff Chain of Solutions.

binary tree, or can the same Markoff number be generated by two different routes through it? There are occasional claims to have proved that the Markoff numbers are unique, but so far the proofs appear to be fallacious.

If $M(N)$ is the number of triples with $x \le y \le z \le N$, Zagier has shown that $M(N) = C(\ln N)^2 + O((\ln N)^{1+\varepsilon})$ where $C \approx 0.180717105$, and calculations lead him to conjecture that $M(N) = C(\ln(3x))^2 + O((\ln x)^{1/2+\varepsilon})$, or equivalently, that the $n$th Markoff number, $m_n$, is $(\frac{1}{3} + O(n^{-1/4+\varepsilon}))A^{\sqrt{n}}$ where $A = e^{1/\sqrt{C}} \approx 10.5101504$. He has no results on distinctness, but can shown that the problem is equivalent to the insolvability of a certain system of diophantine equations.

J. W. S. Cassels, An Introduction to Diophantine Approximation, Cambridge, 1957, 27–44.

H. Cohn, Approach to Markoff's minimal forms through modular functions, *Ann. Math. Princeton* (2) **61** (1955) 1–12.

T. W. Cusick, The largest gaps in the lower Markoff spectrum, *Duke Math. J.* **41** (1974) 453–463; *MR* **57** #5902.

L. E. Dickson, *Studies in the Theory of Numbers*, Chicago Univ. Press, 1930, Chap. VII.

G. Frobenius, Über die Markoffschen Zahlen, *S.-B. Preuss. Akad. Wiss. Berlin* (1913) 458–487.

A. Markoff, Sur les formes quadratiques binaires indéfinies, *Math. Ann.* **15** (1879) 381–409.

R. Remak, Über indefinite binäre quadratische Minimalformen, *Math. Ann.* **92** (1924) 155–182.

R. Remak, Über die geometrische Darstellung der indefiniten binären quadratischen Minimalformen, *Jber. Deutsch Mat.-Verein* **33** (1925) 288–245.

Gerhard Rosenberger, The uniqueness of the Markoff numbers, *Math. Comput.* **30** (1976) 361–365; but see *MR* **53** #280.

L. Ja. Vulah, The diophantine equation $p^2 + 2q^2 + 3r^2 = 6pqr$ and the Markoff spectrum (Russian), *Trudy Moskov. Inst. Radiotehn. Èlektron. i Avtomat. Vyp. 67 Mat.* (1973) 105–112, 152; *MR* **58** #21957.

Don B. Zagier, Distribution of Markov numbers, Abstract 796–A37, *Notices Amer. Math. Soc.*, **26** (1979) A-543.

**D13.** Erdös asked for solutions of the equation $x^x y^y = z^z$, apart from the trivial ones $y = 1$, $x = z$. Chao Ko found an infinity of solutions, of which the first three are

| $x$ | $y$ | $z$ |
|---|---|---|
| $12^6$ | $6^8$ | $2^{11}3^7$ |
| $224^{14}$ | $112^{16}$ | $2^{68}7^{15}$ |
| $61440^{30}$ | $30720^{32}$ | $2^{357}15^{31}$ |

Did he find them all?

Claude Anderson conjectured that the equation $w^w x^x y^y = z^z$ has no solution with $1 < w < x < y$.

Chao Ko, Note on the diophantine equation $x^x y^y = z^z$, *J. Chinese Math. Soc.* **2** (1940) 205–207; *MR* **2**, 346.

W. H. Mills, An unsolved diophantine equation, in *Report Inst. Theory of Numbers*, Boulder, Colorado, 1959, 258–268.

**D14.** Leo Moser asked for integers $a_1, a_2, b_j$ $(1 \le j \le n)$ such that the $2n$ numbers $a_i + b_j$ are all squares. This can be achieved by making $a_2 - a_1$ a sufficiently composite number; for example, $a_1 = 0$, $a_2 = 2^{2n+1}$, $b_j = (2^{2n-j} - 2^{j-1})^2$.

The problem generalizes to $a_i + b_j$ made square, where $1 \le i \le m$. Infinitely many solutions for $m = n = 3$ are found by taking $a_i + b_j$ to be the squares of the numbers in the first three columns of the array

$$\begin{array}{cccc} \frac{1}{2}(ps+qr) & \frac{1}{2}(qs+rp) & \frac{1}{2}(rs+pq) & \frac{1}{2}(pqr+s) \\ \frac{1}{2}(ps-qr) & \frac{1}{2}(qs-rp) & \frac{1}{2}(rs-pq) & \frac{1}{2}(pqr-s) \\ \frac{1}{4}(p^2+s^2-q^2-r^2) & \frac{1}{4}(q^2+s^2-r^2-p^2) & \frac{1}{4}(r^2+s^2-p^2-q^2) & t \end{array}$$

where it's convenient to take $p, q, r, s$ to be odd. This extends to $m = 3$, $n = 4$ if we include the fourth column and can find a nontrivial solution of the diophantine equation

$$16t^2 = (s^2 - p^2 - q^2 - r^2 + 2)^2 + 4(p^2 - 1)(q^2 - 1)(r^2 - 1).$$

One such solution is given by $q = 2p + 1$, $r = 2p - 1$, $t = 2p^3 - p - 1$ with $s$ and $p$ related by the Pell equation

$$17s^2 - (17p - 2)^2 = 72$$

which has an infinity of solutions. If $(s, p) = (21, -5)$ or $(219, -53)$ we have the arrays

$$\begin{array}{cccc} 3^2 & 67^2 & 93^2 & 237^2 \\ 102^2 & 122^2 & 138^2 & 258^2 \\ 66^2 & 94^2 & 114^2 & 246^2 \end{array} \quad \text{or} \quad \begin{array}{cccc} 186^2 & 8662^2 & 8934^2 & 297618^2 \\ 11421^2 & 14333^2 & 14499^2 & 297837^2 \\ 7074^2 & 11182^2 & 11394^2 & 297702^2 \end{array}$$

What about larger values of $n$ and $m$?

**D15.** Erdös and Leo Moser (and see earlier references) also asked the analogous question: are there, for every $n$, $n$ distinct numbers such that the sum of any pair is a square? For $n = 3$ we can take

$$a_1 = \tfrac{1}{2}(q^2 + r^2 - p^2) \qquad a_2 = \tfrac{1}{2}(r^2 + p^2 - q^2) \qquad a_3 = \tfrac{1}{2}(p^2 + q^2 - r^2)$$

and for $n = 4$ we may augment these by taking $s$ to be any number expressible as the sum of two squares in three distinct ways

$$s = u^2 + p^2 = v^2 + q^2 = w^2 + r^2 \quad \text{and} \quad a_4 = s - \tfrac{1}{2}(p^2 + q^2 - r^2).$$

Jean Lagrange has given a quite general parametric solution for $n = 5$ and a simplification of it which appears to give a majority of all solutions. He tabulates the first 80 solutions, calculated by J.-L. Nicolas. The smallest is

$$-4878 \quad 4978 \quad 6903 \quad 12978 \quad 31122$$

and the smallest positive solution (at most one number can be negative) is

$$7442 \quad 28658 \quad 148583 \quad 177458 \quad 763442.$$

In a letter dated 72:05:19 he sends the following solution for $n = 6$:

$$-15863902 \quad 17798783 \quad 21126338 \quad 49064546 \quad 82221218 \quad 447422978$$

In fact the problem goes back to T. Baker who found five integers whose sums in pairs were squares, and C. Gill who found five whose sums in threes were squares.

T. Baker, *The Gentleman's Diary*, or *Math. Repository*, London, 1839, 33–5, Quest. 1385.

C. Gill, *Application of the Angular Analysis to Indeterminate Problems of Degree 2*, N.Y. 1848, p. 60.

J. Lagrange, Cinq nombres dont les sommes deux à deux sont des carrés, *Séminaire Delange-Pisot-Poitou* (*Théorie des nombres*) *12e année*, **20**, 1970–71, 10pp.

Jean Lagrange, Six entiers dont les sommes deux à deux sont des carrés, *Acta Arith.* (to appear).

Jean-Louis Nicolas, 6 nombres dont les sommes deux à deux sont des carrés, *Bull. Soc. Math. France*, Mém No 49–50 (1977) 141–143; *MR* **58** #482.

A. W. Thatcher, A prize problem, *Math. Gaz.* **61** (1977) 64.

**D16.** J. G. Mauldon asks how many different triples of positive integers can you have with the same sum and the same product.

For collections of four such triples he says that the smallest common sum appears to be 118, arising from the triples (14, 50, 54), (15, 40, 63), (18, 30, 70), (21, 25, 72), while the smallest common product appears to be 25200, arising from (6, 56, 75), (7, 40, 90), (9, 28, 100), (12, 20, 105). As an example of an infinite family of such (primitive) triples, he gives

$$(16ka, bc, 15d), \ (10ka, 4bc, 6d), \ (15kb, ad, 16c), \ (6kb, 4ad, 10c)$$

where $a = k + 2$, $b = k + 3$, $c = 2k + 7$, and $d = 3k + 7$. The only example of five triples he had found was (6, 480, 495), (11, 160, 810), (12, 144, 825), (20, 81, 880), and (33, 48, 900).

No example of six such triples is known, although there appears to be no reason why there shouldn't be an arbitrarily large number of triples.

**D17.** Erdös and Selfridge have proved that the product of consecutive integers is never a power, and the binomial coefficient $\binom{n}{k}$ (see B31) is never a power for $n \geq 2k \geq 8$. If $k = 2$, then $\binom{n}{k}$ is a square infinitely often, but Tijdeman's methods (see D9) will probably show that it is never a higher power, and that $k = 3$ never gives a power, apart from $n = 50$ (see D3).

Erdös and Graham ask if the product of two or more disjoint blocks of consecutive integers can be a power. Pomerance has noted that $\prod_{i=1}^{4}(a_i - 1)a_i(a_i + 1)$ is a square if $a_1 = 2^{n-1}$, $a_2 = 2^n$, $a_3 = 2^{2n-1} - 1$, $a_4 = 2^{2n} - 1$, but Erdös and Graham suggest that if $l \geq 4$, then $\prod_{i=1}^{k}\prod_{j=1}^{l}(a_i + j)$ is a square on only a finite number of occasions.

P. Erdös, On consecutive integers, *Nieuw Arch. Wisk.* **3** (1955) 124–128.
P. Erdös and J. L. Selfridge, the product of consecutive integers is never a power, *Illinois J. Math.* **19** (1975) 292–301.

**D18.** Is there a rational box? Our treatment of this notorious unsolved problem is owed almost entirely to John Leech. Does there exist a **perfect cuboid**, with integer edges $x_i$, face diagonals $y_i$ and body diagonal $z$: are there solutions of the simultaneous diophantine equations

$$x_{i+1}^2 + x_{i+2}^2 = y_i^2, \tag{A}$$

$$\sum x_i^2 = z^2 \qquad ? \tag{B}$$

($i = 1$, 2, or 3; where necessary, subscripts are reduced modulo 3.)

Martin Gardner asked if any six of $x_i$, $y_i$, $z$ could be integers. Here there are three problems: just the body diagonal $z$ irrational; just one edge $x_3$ irrational; just one face diagonal $y_1$ irrational.

**Problem 1.** We require solutions to the three equations (A). Suppose such solutions have **generators** $a_i$, $b_i$

$$x_{i+1}:x_{i+2}:y_i = 2a_ib_i:a_i^2 - b_i^2:a_i^2 + b_i^2$$

Then we want integer solutions of

$$\prod \frac{a_i^2 - b_i^2}{2a_ib_i} = 1 \tag{C}$$

We can assume that the generator pairs have opposite parity and replace (C) by

$$\frac{a_1^2 - b_1^2}{2a_1b_1} \cdot \frac{a_2^2 - b_2^2}{2a_2b_2} = \frac{\alpha^2 - \beta^2}{2\alpha\beta} \tag{D}$$

An example is

(E) $$\frac{6^2-5^2}{2\times 6\times 5}\cdot\frac{11^2-2^2}{2\times 11\times 2}=\frac{8^2-5^2}{2\times 8\times 5}$$

Kraitchik gave 241 + 18 − 2 cuboids with odd edge less than a million. Lal and Blundon listed all cuboids obtainable from (D) with $a_1, b_1, \alpha, \beta \le 70$, including the curious pair (1008, 1100, 1155), (1008, 1100, 12075). Leech has deposited a list of all solutions of (D) with two pairs of $a_1, b_1; a_2, b_2; \alpha, \beta \le 376$.

Reversal of the cyclic order of subscripts in (C) leads to the **derived cuboid**: example (E) gives the least solution (240, 44, 117), known to Euler, and the derived cuboid (429, 2340, 880). Note that $240\times 429 = 44\times 2340 = 117\times 880$.

Several parametric solutions are known: the simplest, also known to Euler, is

(F) $$\alpha = 2(p^2-q^2), \qquad a_1 = 4pq, \qquad b_1 = \beta = p^2+q^2$$

For $a_1$, $b_1$ fixed, (D) is equivalent to the plane cubic curve

$$\frac{a_1^2-b_1^2}{2a_1b_1}=\frac{u^2-1}{2u}\cdot\frac{2v}{v^2-1}$$

whose rational points are finitely generated, so Mordell tells us that one solution leads to an infinity. But not all rationals $a_1/b_1$ occur in solutions: $a_1/b_1 = 2$ is impossible, so there is no rational cuboid with a pair of edges in the ratio 3:4.

**Problem 2.** Just an edge irrational. We want $x_1^2 + x_2^2 = y_3^2$ with $t + x_1^2$, $t + x_2^2$, $t + y_3^2$ all squares. This was proposed by "Mahatma" and readers gave $x_1 = 124$, $x_2 = 957$, $t = 13852800$. Bromhead extended tnis to a parametric solution. An infinity of solutions is given by

(G) $$(x_1, x_2, y_3) = 2\xi\eta\zeta(\xi,\eta,\zeta), \qquad t = \zeta^8 - 6\xi^2\eta^2\zeta^4 + \xi^4\eta^4$$

where $(\xi,\eta,\zeta)$ is a Pythagorean triple. The simplest such is $\xi = 5$, $\eta = 12$; $x_1 = 7800$, $x_2 = 18720$; $t = 211773121$.

These are not all. We seek solutions of

(H) $$x_1^2 + x_2^2 = y_3^2, \qquad z^2 = x_1^2 + y_1^2 = x_2^2 + y_2^2$$

other than $z = y_3$ $(t = 0)$. Leech found 100 primitive solutions with $z < 10^5$, 46 of which had $t > 0$. The generators for (H) satisfy

(I) $$\frac{\alpha_1^2+\beta_1^2}{2\alpha_1\beta_1}\cdot\frac{2\alpha_2\beta_2}{\alpha_2^2+\beta_2^2}=\frac{a^2-b^2}{2ab}$$

so solutions for fixed $x_2/x_1$ correspond to rational points on the cubic curve

(J) $$x_1v(u^2+1) = x_2u(v^2+1)$$

The trivial solution $t = 0$ is an **ordinary point** which generates an infinity of solutions for each ratio $x_2/x_1$. Many solutions form cycles of four

$$\text{(K)} \quad \begin{gathered} \zeta^2 = \xi_1^2 + \eta_1^2 = \xi_2^2 + \eta_2^2 = \xi_3^2 + \eta_3^2 = \xi_4^2 + \eta_4^2, \qquad \xi_1\xi_3 = \xi_2\xi_4, \\ \xi_1^2 + \xi_2^2, \quad \xi_2^2 + \xi_3^2, \quad \xi_3^2 + \xi_4^2, \quad \xi_4^2 + \xi_1^2 \quad \text{all squares,} \end{gathered}$$

corresponding to two pairs $x_2/x_1$ which correspond to two points collinear with the point $t = 0$ on (J). Conversely, such a pair of points corresponds to a cycle of four solutions.

**Problem 3.** Just one face diagonal irrational. There are two closely related problems: find three integers whose sums and differences are all squares; find three squares whose differences are squares. The cuboid form of the problem asks for integers satisfying

$$\text{(L)} \qquad x_2^2 + y_2^2 = z^2, \qquad x_1^2 + x_3^2 = y_2^2, \qquad x_1^2 + x_2^2 = y_3^2$$

whose generators satisfy

$$\text{(M)} \qquad \frac{\alpha_1^2 - \beta_1^2}{2\alpha_1\beta_1} \cdot \frac{\alpha_3^2 - \beta_3^2}{2\alpha_3\beta_3} = \frac{\alpha_2^2 + \beta_2^2}{2\alpha_2\beta_2}.$$

Write $u_i = (\alpha_i^2 - \beta_i^2)^2/4\alpha_i^2\beta_i^2$ and (M) becomes $u_1u_3 = 1 + u_2$, an equation investigated by many in the contexts of **cycles** and **frieze patterns**. Solutions occur in cycles of five! Leech listed 35 such with $\alpha_1, \beta_1, \alpha_2, \beta_2 \leq 50$ and deposited a list of all cycles with two pairs $\alpha_i, \beta_i \leq 376$. There is a close connexion with Napier's rules and the construction of rational spherical triangles.

Solutions to (L) are given by $x_2^2 = z^2 - y_2^2 = (p^2 - q^2)(r^2 - s^2)$, $x_3^2 = z^2 - y_3^2 = 4pqrs$, when the products of the numerators and denominators of $(p^2 - q^2)/2pq$ and $(r^2 - s^2)/2rs$ are each squares. Euler made $p, q, r, s$ squares and found differences of fourth powers, e.g., $3^4 - 2^4$, $9^4 - 7^4$, $11^4 - 2^4$, whose products are squares. The first two give the second smallest solution $(117, 520, 756)$ of this type, whose cycle includes the smallest $(104, 153, 672)$, also known to Euler.

We can also express $z^2 = x_2^2 + y_2^2 = x_3^2 + y_3^2$ as the sum of two squares in two different ways: $x_3/x_1 = (\alpha_2^2 - \beta_2^2)/2\alpha_2\beta_2$, $x_2/x_1 = (\alpha_3^2 - \beta_3^2)/2\alpha_3\beta_3$ gives $z^2 = 4(\alpha_2^2\alpha_3^2 + \beta_2^2\beta_3^2)(\alpha_2^2\beta_3^2 + \alpha_3^2\beta_2^2)$. Euler made each factor square and found two rational right triangles of equal area $\frac{1}{2}\alpha_2\alpha_3\beta_2\beta_3$. Diophantus solved this with $\beta_2/\alpha_2 = (s + t)/2r$, $\beta_3/\alpha_3 = s/t$, where $r^2 = s^2 + st + t^2$, $s = l^2 - m^2$, $t = m^2 - n^2$. Put $(l, m, n) = (1, 2, -3)$ and we have a cycle containing the third, fourth, and fifth smallest of these cuboids. Leech found 89 with $z < 10^5$.

For fixed $\alpha_1/\beta_1$ a nontrivial solution to (M) corresponds to an ordinary point on $(\alpha_1^2 - \beta_1^2)u(v^2 - 1) = 2\alpha_1\beta_1 v(u^2 + 1)$ which generates an infinity of solutions. The tangent at the point generates a cycle of special interest.

(M), like (D), but unlike (I), does not have nontrivial solutions for all ratios $\alpha/\beta$; e.g., there are none with $\alpha/\beta = 2$ or 3 and again no cuboids with edges in the ratios 3:4. Here we do not have "derived" cuboids.

Two other parametric forms for ratios $(p^2 - q^2)/2pq$ whose product and quotient are squares, are

$$p = 2m^2 \pm n^2, \qquad r = m^2 \pm 2n^2, \qquad q = s = m^2 \mp n^2.$$

**Four squares whose sums in pairs are square.** A solution of (C) gives three such squares; it may be portrayed as a trivalent **vertex** of a graph, the three edges joining it to **nodes** representing generator pairs for rational triangles. If such a pair occurs in one solution it occurs in infinitely many, so the valence of a *node* is infinite. We seek a subgraph homeomorphic to a tetrahedron $K_4$ whose vertices give four solutions of (C) and whose edges contain nodes corresponding to generator pairs common to pairs of solutions of (C). Lists of solutions have revealed no such subgraph; indeed, not even a closed circuit! Until a circuit is encountered, we need not distinguish between the pairs $a$, $b$ and $a \pm b$.

So no example of four such squares is known. Construction of four squares with five square sums of pairs is straightforward.

**Four squares whose differences are square.** This problem extends (M) analogously to the above extension of (C). A *vertex* is now *pentavalent*, adjacent to a cycle of five nodes or generator pairs. The cyclic order of the edges is important, but not the sense of rotation. A solution of (M) corresponds to three *consecutive* edges, and here we must distinguish between $\alpha$, $\beta$ and $\alpha \pm \beta$: the corresponding nodes are joined with double edges in Figure 11. The nodes are again of infinite valence. Four squares of the required type would correspond to a subgraph homeomorphic to $K_6$, with six vertices and 15 nodes, one on each edge. The only cycle so far found is shown in Figure 11 and this will *not* serve as part of such a subgraph. Although

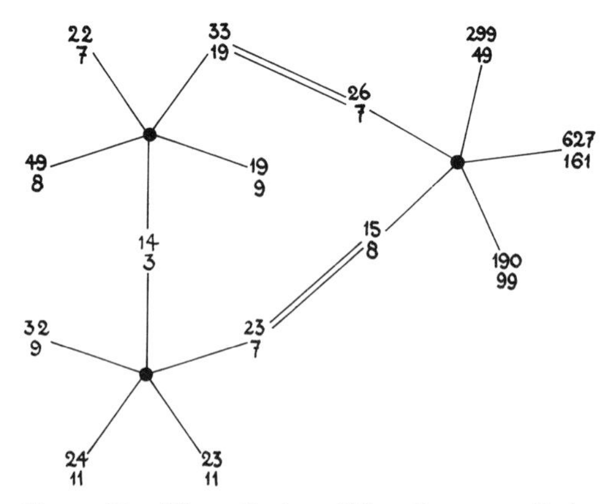


Figure 11. Three Cycles of Five Generator Pairs.

a solution is unlikely, there do not appear to be any congruence conditions which forbid it.

Though there is now no necessary connexion between the generator pairs $\alpha$, $\beta$ and $\alpha \pm \beta$, solutions of (M) containing both pairs do occur. Such a solution leads to a sequence of four squares with sums of 2 or 3 *consecutive* terms all square. How long can such a sequence be? For more than four, we need a sequence of 5-cycles of solutions of (M) each containing adjacent generator pairs $\alpha, \beta; \bar{\alpha}, \bar{\beta}$ where $\alpha \pm \beta$, $\bar{\alpha} \pm \bar{\beta}$ belong one each to the neighboring cycles. Leech found the sequence

$$(56, 31)(17, 6); (23, 11)(23, 7); (15, 8)(26, 7);$$
$$(33, 19)(77, 19); (48, 29)(35, 4); (39, 31)(13, 9)$$

where 23, $11 = 17 \pm 6$, etc. which gives a sequence of eight such squares. The squares of the edges of a perfect cuboid would form an infinite (periodic) sequence. Four nonzero squares with differences all square would lead to a sequence with terms three apart in constant ratio: an integer ratio would give an infinite sequence.

**The perfect rational cuboid.** None of the known numerical solutions to problems 1, 2, and 3 gives a perfect cuboid, and many parametric solutions, for example (G), can be shown not to yield one. Spohn used Pocklington's work to show that *one* of the two mutually derived cuboids of (F) is not perfect and Jean Lagrange has shown that the derived cuboid is never perfect. On the other hand, no known parametric solution is complete so impossibility can't be proved from these alone. A solution of the problem in the previous section need not lead to a perfect cuboid.

**Unsolved problems.** Do three cycles of solutions of (M) exist whose graph is as in Figure 12? Are there ratios $(p^2 - q^2)/2pq$, $(r^2 + s^2)/2rs$ with product

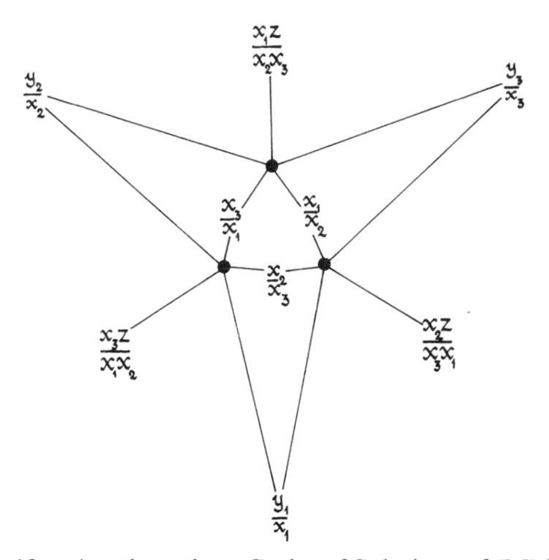


Figure 12. Are there three Cycles of Solutions of (M) like this?

and quotient both of the form $(m^2 - n^2)/2mn$? Is there a nontrivial solution of $(a^2c^2 - b^2d^2)(a^2d^2 - b^2c^2) = (a^2b^2 - c^2d^2)^2$? Such a solution would lead to a perfect cuboid. Is there a 5-cycle of solutions of (M) with $\alpha_1/\beta_1 = (p^2 - q^2)/2pq$, $\alpha_2/\beta_2 = (r^2 + s^2)/2rs$? What circuits, if any, occur in the graph of solutions of (C)? What circuits occur in the graph of solutions of (M)? Are there cycles of solutions of (I) other than those of form (K)? What ratios besides 3/4 can*not* occur as ratios of edges of cuboids in problem 1? In problem 3? Is there a parallelepiped with all edges, face diagonals, and body diagonals rational?

T. Bromhead, On square sums of squares, *Math. Gaz.* **44** (1960) 219–220; *MR* **23** #A1594.

W. S. Burnside, Note on the symmetric group, *Messenger of Math.* **30** (1900–01) 148–153; *J'buch* **32**, 141–142.

E. Z. Chein, On the derived cuboid of an Eulerian triple, *Canad. Math. Bull.* **20** (1977) 509–510; *MR* **57** #12375.

J. H. Conway and H. S. M. Coxeter, Triangulated polygons and frieze patterns, *Math. Gaz.* **57** (1973) 87–94, 175–183 and refs. on pp. 93–94.

H. S. M. Coxeter, Frieze patterns, *Acta Arith.* **18** (1971) 297–310; *MR* **44** #3980.

L. E. Dickson, *History of the Theory of Numbers*, Vol. 2, Diophantine Analysis, Washington, 1920: ch. 15, ref. 28, p. 448 and cross-refs. to pp. 446–458; ch. 19 refs 1–30, 40–45, pp. 497–502, 505–507.

Martin Gardner, Mathematical Games, *Scientific Amer.* **223** #1 (Jul. 1970) 118; correction. *ibid.* #3 (Sep. 1970) 218.

W. Howard Joint, Cycles, Note 1767, *Math. Gaz.* **28** (1944) 196–197.

Maurice Kraitchik, On certain rational cuboids, *Scripta Math.* **11** (1945) 317–326; *MR* **8**, 6.

Maurice Kraitchik, *Théorie des Nombres*, t.3, Analyse Diophantine et applications aux cuboides rationnels, Paris, 1947.

Maurice Kraitchik, Sur les cuboides rationnels, in *Proc. Internat. Congr. Math.* 1954, Vol. 2, Amsterdam, 1954, 33–34.

M. Lal and W. J. Blundon, Solutions of the Diophantine equation $x^2 + y^2 = l^2$, $y^2 + z^2 = m^2$, $z^2 + x^2 = n^2$, *Math. Comput.* **20** (1966) 144–147; *MR* **32** #4082.

J. Lagrange, Sur le dérivé du cuboïde eulerien, *Canad. Math. Bull.* **22** (1979) 239–241; *MR* **80h**:10022.

J. Leech, The location of four squares in an arithmetic progression with some applications, in *Computers in Number Theory*, Academic Press, London, 1971, 83–98; *MR* **47** #4913.

J. Leech, The rational cuboid revisited, *Amer. Math. Monthly* **84** (1977) 518–533. Corrections (Jean Lagrange) **85** (1978) 473; *MR* **58** #16492.

J. Leech, Five tables related to rational cuboids, *Math. Comput.* **32** (1978) 657–659.

R. C. Lyness, Cycles, Note 1581, *Math. Gaz.* **26** (1942) 62; Note 1847 *ibid* **29** (1945) 231–233; Note 2952, *ibid* **45** (1961) 207–209.

"Mahatma", Problem 78, *The A.M.A.* (*J. Assist. Masters Assoc. London*) **44** (1949) 188. Solutions J. Hancock, J. Peacock, N. A. Phillips, *ibid.* 225.

Eliakim Hastings Moore, The cross-ratio group of $n!$ Cremona-transformations of order $n - 3$ in flat space of $n - 3$ dimensions, *Amer. J. Math.* **30** (1900) 279–291; *J'buch* **31**, 655.

L. J. Mordell, On the rational solutions of the indeterminate equations of the third and fourth degrees, *Proc. Cambridge Philos. Soc.* **21** (1922) 179–192.

H. C. Pocklington, Some diophantine impossibilities, *Proc. Cambridge Philos. Soc.* **17** (1914) 110–121, esp. p. 116.

W. W. Sawyer, Lyness's periodic sequence, Note 2951, *Math. Gaz.* **45** (1961) 207.
W. Sierpiński, *A Selection of Problems in the Theory of Numbers*, Pergamon, Oxford, 1964, p. 112.
W. G. Spohn, On the integral cuboid, *Amer. Math. Monthly*, **79** (1972) 57–59; *MR* **46** #7158.
W. G. Spohn, On the derived cuboid, *Canad. Math. Bull.* **17** (1974) 575–577; *MR* **51** #12693.

**D19.** Is there a point all of whose distances from the corners of the unit square are rational? It was earlier thought that there might not be any nontrivial example (i.e., an example not on the side of the square) of a point with *three* such rational distances, but John Conway and Mike Guy found an infinity of integer solutions of

$$(s^2 + b^2 - a^2)^2 + (s^2 + b^2 - c^2)^2 = (2bs)^2$$

where $a$, $b$, $c$ are the distances of a point from three corners of a square of side $s$. There are relations between such solutions as shown in Figure 13.

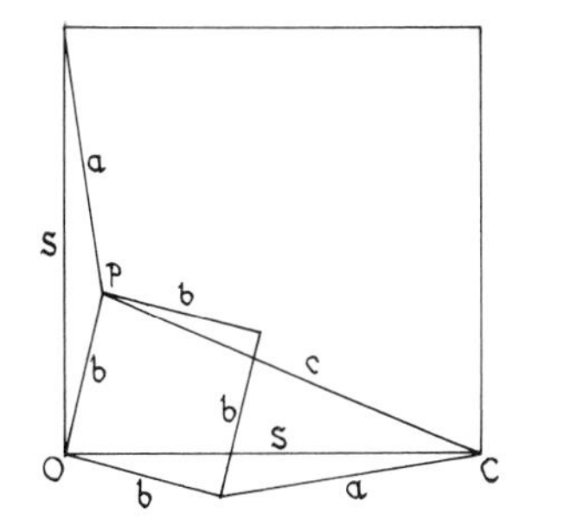


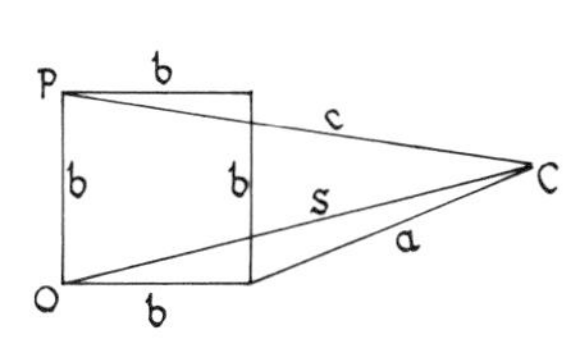


Figure 13. Two Related Solutions of the Three Distance Problem.

For the fourth distance $d$ to be integral we also need $a^2 + c^2 = b^2 + d^2$. In the three distance problem, one of $s$, $a$, $b$, $c$ is divisible by 3, one by 4, and one by 5. In the four distance problem, $s$ is a multiple of 4 and $a$, $b$, $c$, $d$ are odd (assuming that there is no common factor). If $s$ is not a multiple of 3 (respectively 5) then two of $a$, $b$, $c$, $d$ are divisible by 3 (respectively 5).

There are infinitely many solutions of the corresponding problem of integer distances $a$, $b$, $c$ from the corners of an equilateral triangle of side $t$. In each of these one of $a$, $b$, $c$, $t$ is divisible by 3, one by 5, one by 7, and one by 8.

Jerry Bergum asks for what integers $n$ do there exist positive integers $x$, $y$ with $(x, y) = 1$, $x$ even, and $x^2 + y^2 = b^2$ and $x^2 + (y - nx)^2 = c^2$ both perfect squares. If $n = 2m(2m^2 + 1)$, then $x = 4m(4m^2 + 1)$, $y = mx + 1$ is a solution. There are no solutions for $n = \pm 1$, $\pm 2$, $\pm 3$, or $\pm 4$. If $n = 8$, the least $x$ for which there is a $y$ is $x = 2996760 = 2^3 \times 3 \times 5 \times 13 \times 17 \times 113$. The connexion between this problem and the original one is that $(x, y)$ are the coordinates of $P$ at distances $b$ and $c$ from the origin $O$ and an adjacent corner $C$ of the square of side $s = nx$ where $n$ is an integer.

Ron Evans notes that the problem may be stated: which integers $n$ occur as the ratios base/height in integer-sided triangles? The sign of $n$ is $\pm$ according as the triangle is acute or obtuse (e.g., $n = -29$, $x = 120$, $y = 119$ is a solution). He also asks the dual problem: find every integer-sided triangle whose base divides its height. Here the height/base ratios 1 and 2 can't occur, but 3 can (e.g., base 4; sides 13, 15; height 12).

R. B. Eggleton, Tiling the plane with triangles, *Discrete Math.* **7** (1974) 53–65.
R. B. Eggleton, Where do all the triangles go? *Amer. Math. Monthly* **82** (1975) 499–501.
Ronald Evans, Problem E2685, *Amer. Math. Monthly* **84** (1977) 820.
N. J. Fine, On rational triangles, *Amer. Math. Monthly* **83** (1976) 517–521.
J. G. Mauldon, An impossible triangle, *Amer. Math. Monthly* **86** (1979) 785–786.
C. Pomerance, On a tiling problem of R. B. Eggleton, *Discrete Math.* **18** (1977) 63–70.

**D20.** Are there six points in the plane, no three in line, no four on a circle, all of whose mutual distances are rational? A countable infinity of (noncollinear) points having mutual distances which are rational is known to exist, but all except two lie on a circle.

There are two opposite extreme conjectures: (a) that there is a number $c$ such that any $n$ points whose mutual distances are rational must contain at least $n - c$ which lie on a circle or straight line, and (b) Besicovitch's conjecture that any polygon can be approximated arbitrarily closely by a rational polygon.

If we relax the conditions to, say, at most 4 points on a line and at most 4 points on a circle, then John Leech has found infinitely many sets of 7 points of the type illustrated in Figure 14(a) and the set of 8 shown in Figure 14(b). He says that the latter seems to be a numerical freak resulting from solving "too many" equations (four in three homogeneous variables).

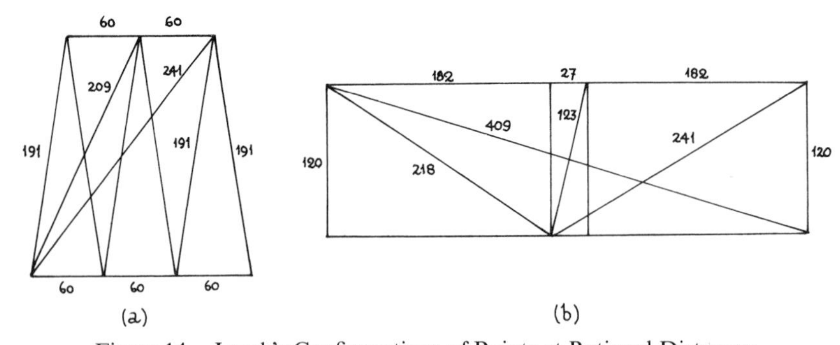


Figure 14. Leech's Configurations of Points at Rational Distances.

J. H. J. Almering, Rational quadrilaterals, *Nederl. Akad. Wetensch. Proc. Ser. A* **66** = Indag. Math. **25** (1963) 192–199; II, ibid. **68** = **27** (1965) 290–304; *MR* **26** #4963, **31** #3375.
D. D. Ang, D. E. Daykin and T. K. Sheng, On Schoenberg's rational polygon problem, *J. Austral. Math. Soc.* **9** (1969) 337–344; *MR* **39** #6816.

A. S. Besicovitch, Rational polygons, *Mathematika* **6** (1959) 98; *MR* **22** #1557.
D. E. Daykin, Rational polygons, *Mathematika* **10** (1963) 125–131; *MR* **30** #63.
D. E. Daykin, Rational triangles and parallelograms, *Math. Mag.* **38** (1965) 46–47.
L. J. Mordell, Rational quadrilaterals, *J. London Math. Soc.* **35** (1960) 277–282.
T. K. Sheng, Rational polygons, *J. Austral. Math. Soc.* **6** (1966) 452–459; *MR* **35** #137.
T. K. Sheng and D. E. Daykin, On approximating polygons by rational polygons, *Math. Mag.* **38** (1966) 299–300; *MR* **34** #7463.

**D21.** Is there a triangle with integer sides, medians, and area? There are, in the literature, incorrect "proofs" of impossibility, but the problem remains open.

**D22.** Are there simplexes in any number of dimensions, all of whose contents (lengths, areas, volumes, hypervolumes) are rational? The answer is "yes" in two dimensions; there are infinitely many **Heron triangles** with rational sides and area. An example is a triangle of sides 13, 14, 15 which has area 84. If the answer is "yes" in three dimensions, can all tetrahedra be approximated arbitrarily closely by such rational ones?

**D23.** Another of many unsolved diophantine equations is

$$(x^2 - 1)(y^2 - 1) = (z^2 - 1)^2$$

though Schinzel and Sierpiński have found all solutions for which $x - y = 2z$.

A. Schinzel and W. Sierpiński, Sur l'equation diophantienne $(x^2 - 1)(y^2 - 1) = [((y - x)/2)^2 - 1]^2$, *Elem. Math.* **18** (1963) 132–133; *MR* **29** #1180.

**D24.** For $k > 2$, the equation $a_1 a_2 \cdots a_k = a_1 + a_2 + \cdots + a_k$ has the solution $a_1 = 2$, $a_2 = k$, $a_3 = a_4 = \cdots = a_k = 1$. Schinzel showed that there is no other solution for $k = 6$ or $k = 24$. Misiurewicz has shown that $k = 2$, 3, 4, 6, 24, 144, 174, and 444 are the only $k < 1000$ for which there is exactly one solution.

M. Misiurewicz, Ungelöste Probleme, *Elem. Math.* **21** (1966) 90.

**D25.** Are the only solutions of $n! + 1 = x^2$ given by $n = 4$, 5 and 7? Erdös and Obláth settled the case $n! = x^p \pm y^p$ with $(x, y) = 1$ and $p > 2$, but the case $p = 2$ is still open.

Simmons notes that $n! = (m - 1)m(m + 1)$ for $(m, n) = (2, 3)$, $(3, 4)$, $(5, 5)$ and $(9, 6)$ and asks if there are other solutions.

H. Brocard, Question 1532, *Nouv. Corresp. Math.* **2** (1876) 287; *Nouv. Ann. Math.* (3) **4** (1885) 391.
P. Erdös and R. Obláth, Über diophantische Gleichungen der Form $n! = x^p \pm y^p$ und $n! \pm m! = x^p$, *Acta Szeged* **8** (1937) 241–255.

M. Kraitchik, Recherches sur la Théorie des Nombres, t. 1, Gauthier-Villars, Paris, 1924, 38–41.
Richard M. Pollack and Harold N. Shapiro, The next to last case of a factorial diophantine equation, *Comm. Pure Appl. Math.* **26** (1973) 313–325; *MR* **50** #12915.
G. J. Simmons, A factorial conjecture, *J. Recreational Math.* **1** (1968) 38.

**D26.** Stark asks which Fibonacci numbers (see A3) are half the difference (or sum) of two cubes. This is related to the problem of finding all complex quadratic fields of class-number 2. Examples: $1 = \frac{1}{2}(1^3 + 1^3)$, $8 = \frac{1}{2}(2^3 + 2^3)$, $13 = \frac{1}{2}(3^3 - 1^3)$.

Vern Hoggatt asks if 1, 3, 21, and 55 are the only Fibonacci numbers which are triangular (i.e., of the form $\frac{1}{2}m(m+1)$).

H. M. Stark, Problem 23, *Summer Institute on Number Theory*, Stony Brook, 1969.

**D27.** **Congruent numbers**, which are related to pythagorean triangles, have an ancient history, several examples (5, 6, 14, the seventeen entries CA in Table 7, and ten more $>1000$) being given in an Arab manuscript more than a thousand years ago. On the other hand, they still appear to be of current interest. They are those integers $a$ for which

$$x^2 + ay^2 = z^2, \qquad x^2 - ay^2 = t^2$$

have simultaneous integer solutions. Perhaps part of the fascination is the often inordinate size of smallest solutions. For example, $a = 101$ is a congruent number and Bastien gave the smallest solution:

$$x = 2015\ 242462\ 949760\ 001961 \qquad y = 118\ 171431\ 852779\ 451900$$
$$z = 2339\ 148435\ 306225\ 006961 \qquad t = 1628\ 124370\ 727269\ 996961.$$

Equivalently, congruent numbers are those $a$ for which there are solutions of the diophantine equation

$$x^4 - a^2y^4 = u^2.$$

Dickson's *History* gives many early references, including Leonardo of Pisa (Fibonacci); Genocchi; and Gérardin, who gave 7, 22, 41, 69, 77, the twenty Arabic examples and the forty-three entries CG in Table 7. It is clear that we need consider only squarefree values of $a$ since if $b = ad^2$, then a solution $(x, y, z, t)$ of the original equations corresponds to $(dx, y, dz, dt)$ of the equations with $b$ in place of $a$.

It has been conjectured that squarefree numbers $\equiv 5$, 6, or 7 (mod 8) are congruent. Nelson Stephens showed that this follows from the Selmer conjecture for elliptic curves (see Cassels) so that it is known from the work of Heegner (see Birch) that this is true for primes $\equiv 5$ or 7 (mod 8) and the doubles of primes $\equiv 3$ (mod 8). These are the entries C5, C7, and C6 in Table 7. Bastien observed that the following are noncongruent: primes

≡3 (mod 8); products of two such primes; the doubles of primes ≡5 (mod 8); the doubles of the products of two such primes; and the doubles of primes ≡9 (mod 16); these are the entries N3, N9, NX (for $10 = 2 \times 5$), NL (for $50 = 2 \times 5 \times 5$) and N2 in Table 7. He gave some other noncongruent numbers (NB in Table 7) and stated that $a$ was noncongruent if it was a prime ≡1 (mod 8) with $a = b^2 + c^2$ and $b + c$ a nonresidue of $a$ (see F7). This accounts for several of the entries N1. So that the table can also serve as a table of primes <1000, those ≡1 (mod 8) are all indicated (by N1, C1, or just 1, if the status is unknown). Entries C& and N& are taken from Alter et al, and entries CJ and NJ are from Jean Lagrange's remarkable thesis. A square (□) indicates that the number contains a repeated factor and a blank that the status is unknown to the present writer.

Table 7. Known Congruent (C) and Noncongruent (N) Numbers less than 1000. The entry for $a = 40c + r$ is in column c and row r.

| r \ c = | 0 | 1 | 2 | 3 | 4 | 5 | 6 | 7 | 8 | 9 | 10 | 11 | 12 | 13 | 14 | 15 | 16 | 17 | 18 | 19 | 20 | 21 | 22 | 23 | 24 = c | r |
|---|---|---|---|---|---|---|---|---|---|---|---|---|---|---|---|---|---|---|---|---|---|---|---|---|---|---|
| 1 | NB | C1 | □ | □ | CG | N9 | N1 | N1 | N9 | □ | N1 | □ | NJ | N1 | CG | N1 | N1 | N9 | C& | C1 | □ | □ | 1 | N9 | □ | 1 |
| 2 | NB | NB | N2 | NX | □ | NX | □ |  | NJ | NX | NJ | CG |  | □ | N2 | CG | NJ | NJ | □ | NJ |  | NX | □ | NX | NL | 2 |
| 3 | N3 | N3 | N3 | N& | N3 | NJ | □ | N3 | C& | □ | NJ | N3 | NJ | N3 | N3 | □ | N3 | N3 | CJ | NJ | NJ | NJ | N3 | NJ | □ | 3 |
| 4 | □ | □ | □ | □ | □ | □ | □ | □ | □ | □ | □ | □ | □ | □ | □ | □ | □ | □ | □ | □ | □ | □ | □ | □ | □ | 4 |
| 5 | C5 | □ | CG | □ | CG | CG | □ | C& | □ | CJ | □ | C& | CJ | □ | C& | □ | C& | CJ | □ | □ | C& | □ | CJ | □ |  | 5 |
| 6 | C6 | C6 | C6 | □ | C6 | C6 | C& | CA | C6 | C& | CJ | C6 | □ | C6 | C6 | CJ | CG | □ | □ | C6 | CG | □ | C6 | C6 | CG | 6 |
| 7 | C7 | C7 | CG | C7 | C7 | □ | CJ | C& | CJ | C7 | CJ | CJ | C7 | C& | □ | C7 | C7 | CJ | C7 | CJ | CJ | □ | C7 | □ | C7 | 7 |
| 8 | □ | □ | □ | □ | □ | □ | □ | □ | □ | □ | □ | □ | □ | □ | □ | □ | □ | □ | □ | □ | □ | □ | □ | □ | □ | 8 |
| 9 | □ | □ | N1 | N9 | □ | N9 | N9 | □ | NJ | □ | N1 | N1 | N9 | □ | 1 | C& | N9 | C& | □ | N1 | 1 | N9 | C& | N1 | NJ | 9 |
| 10 | NX | □ | □ | NL | N& | CA | □ | NL | CA | NL | CG | □ | □ | NL | NJ | NL | □ | NJ | NJ | NJ | □ | □ | CG | NJ | NJ | 10 |
| 11 | N3 | NB | NB | N3 | □ | N3 | N3 | CG | N3 | CG | NJ | NJ | N3 | □ | N3 | NJ | CG | N3 | C& | NJ | N3 | NJ | □ | □ | N3 | 11 |
| 12 | □ | □ | □ | □ | □ | □ | □ | □ | □ | □ | □ | □ | □ | □ | □ | □ | □ | □ | □ | □ | □ | □ | □ | □ | □ | 12 |
| 13 | C5 | C5 | CG | CJ | C5 | CJ | CJ | C5 | □ | C5 | CJ | CJ | CJ | C& |  | C5 | C5 | □ | C5 | C5 | C& | C5 |  |  | CJ | 13 |
| 14 | C6 | □ | C6 | C6 | CG | C6 | C6 | □ | C6 | CJ | □ | C6 | CJ | CJ | C& | C6 | CJ | C6 | C6 | □ | C& | CJ | CJ | C6 | C& | 14 |
| 15 | CA | CG | CG | □ | □ | C& | CG | CJ | CJ | □ | CJ | C& | □ | C& | □ | C& | CJ | CJ | □ | □ | CJ | □ | CJ | C& | □ | 15 |
| 16 | □ | □ | □ | □ | □ | □ | □ | □ | □ | □ | □ | □ | □ | □ | □ | □ | □ | □ | □ | □ | □ | □ | □ | □ | □ | 16 |
| 17 | N1 | N9 | N1 | C1 | N9 | NJ | C1 | □ | N1 | NJ | N9 | C1 | NJ | N9 | 1 | N1 | □ | NJ | N9 | C& | N9 | 1 |  | N1 | N1 | 17 |
| 18 | □ | NX | □ | CG | N2 | NX | NJ | NX | □ | □ | NJ | NX | NJ | NX | □ | NJ | C& | NX | □ | NX | N2 | NJ |  |  | NJ | 18 |
| 19 | N3 | N3 | □ | N3 | N3 | C& | NJ | CG | NJ | N3 | N3 | □ | N3 | □ | NJ | N3 | N3 | NJ | N3 | NJ | □ | N3 | NJ |  | NJ | 19 |
| 20 | □ | □ | □ | □ | □ | □ | □ | □ | □ | □ | □ | □ | □ | □ | □ | □ | □ | □ | □ | □ | □ | □ | □ | □ | □ | 20 |
| 21 | CA | C5 | C5 | C& | C5 | CA | □ | CJ | CJ | CJ | C5 | C5 | CJ | C5 | CJ | □ | C5 | C5 | CG | CJ | C5 | CJ | C& | C5 | □ | 21 |
| 22 | C6 | C6 | CG | C6 | C& | CJ | C6 | C6 | □ | C6 | C6 | CG | C6 | C6 | C& | C6 | C6 | □ | CJ | CJ | CJ | C6 | CJ | CJ | C6 | 22 |
| 23 | C7 | □ | C7 | C& | CJ | C7 | C7 | CJ | □ | C7 | □ | C7 | C7 |  | CG |  | C& | CJ | C7 | □ | C7 | C7 | C& | C& | C7 | 23 |
| 24 | □ | □ | □ | □ | □ | □ | □ | □ | □ | □ | □ | □ | □ | □ | □ | □ | □ | □ | □ | □ | □ | □ | □ | □ | □ | 24 |
| 25 | □ | CA | NJ | CG | NJ | □ | CG | NJ | NJ | NJ | □ | CG | C& | NJ | □ | □ | NJ | NJ | NJ | NJ | □ | NJ | C& | □ | C& | 25 |
| 26 | NX | NB | NX | N2 | N& | C& | NJ | □ | NX | C& | C& | N2 | NJ | CA | NX | N2 | □ |  | NX | NJ | NJ | C& | NJ | NJ | NJ | 26 |
| 27 | □ | N3 | N3 | □ | N& | N3 | NJ | N3 | N3 | □ | NJ | N3 | □ | N3 | N3 |  | NJ | NJ | □ | N3 | N3 | □ | N3 | N3 | C& | 27 |
| 28 | □ | □ | □ | □ | □ | □ | □ | □ | □ | □ | □ | □ | □ | □ | □ | □ | □ | □ | □ | □ | □ | □ | □ | □ | □ | 28 |
| 29 | C5 | CG | C5 | C5 | □ | C5 | C5 | CJ | C5 | C5 | CA | CJ | C5 | □ | CJ | C& | C& | C5 | CJ | CJ | C5 | CJ | □ | C& | CJ | 29 |
| 30 | CA | CA | CA | □ | CA | CJ | □ | CG | □ | CA | CJ | CG | CG | □ | CJ | □ | C& | C& | □ | CJ | CJ | C& | C& | □ | □ | 30 |
| 31 | C7 | C7 | CG | C7 | C7 | CA | C7 | C7 | □ | C& | C7 | CJ | C& | CJ | CJ | C7 | C& | □ | C7 | C& | CJ | CJ | C7 | C& | C7 | 31 |
| 32 | □ | □ | □ | □ | □ | □ | □ | □ | □ | □ | □ | □ | □ | □ | □ | □ | □ | □ | □ | □ | □ | □ | □ | □ | □ | 32 |
| 33 | N9 | N1 | N1 | □ | N1 | N1 | NJ | C1 | C1 | N9 | N1 | N9 | □ | NJ | 1 | N9 | N1 | NJ | N9 | C& | □ | □ | N9 | 1 | N9 | 33 |
| 34 | CA | NX | N& | CA | C& | □ | N2 | NX | NJ | NX | CG | NJ | C& | NX | □ | NX | C& | NJ | NL | NX | NJ | NJ | N2 | □ | NJ | 34 |
| 35 | NB | □ | N& | NJ | NJ | NJ | □ | □ | NJ | C& | NJ | □ | NJ | NJ | NJ | NJ | □ | NJ | NJ | NJ | NJ | □ | C& | NJ | C& | 35 |
| 36 | □ | □ | □ | □ | □ | □ | □ | □ | □ | □ | □ | □ | □ | □ | □ | □ | □ | □ | □ | □ | □ | □ | □ | □ | □ | 36 |
| 37 | C5 | CG | □ | C5 | C5 | CJ | C5 | C5 | CG | C5 | CJ | □ | CG | C5 |  | □ | C5 |  | C5 | C5 | □ | C5 |  | C& | C5 | 37 |
| 38 | C6 | CG | C6 | C6 | □ | CJ | C6 | C& | C6 | C6 | CG | C6 | C& | □ | CJ | CJ | CJ | C6 | C6 | CG | C6 | C6 | □ | C6 | C6 | 38 |
| 39 | CG | C7 | C& | C& | C7 | C7 | □ | C& | C7 | C& | C7 | C7 | CJ | CJ | C7 | □ | CJ | C7 | CG | C& | C7 | C& | C7 | C& | □ | 39 |
| 40 | □ | □ | □ | □ | □ | □ | □ | □ | □ | □ | □ | □ | □ | □ | □ | □ | □ | □ | □ | □ | □ | □ | □ | □ | □ | 40 |
| r \ c = | 0 | 1 | 2 | 3 | 4 | 5 | 6 | 7 | 8 | 9 | 10 | 11 | 12 | 13 | 14 | 15 | 16 | 17 | 18 | 19 | 20 | 21 | 22 | 23 | 24 = c | r |

It is hard to be sure that one has collected all results, since, especially if they have been obtained constructively, they are not always published. I am indebted to J. A. H. (*Fun with Figures*) Hunter, who has obtained or stimulated several such constructions, for the following items. M. R. Buckley and K. Gallyas (*Fibonacci Assoc. Newsletter*, Sept 1975) discovered that

$$134\ 130664\ 938047\ 228374\ 702001\ 079697^2 \pm 103 \times 7\ 188661\ 768365\ 914788\ 447417\ 161240^2$$

are both squares, while Hunter and Buckley found the minimal representations

$$764646\ 440211\ 958998\ 267241^2 \pm 229 \times 9404\ 506457\ 489780\ 613180^2 = 777777\ 618847\ 556210\ 645041^2 \quad \text{or} \quad 751285\ 786287\ 393798\ 649441^2$$

and $407\ 893921^2 \pm 239 \times 4826640^2 = 414\ 662879^2$ or $401\ 010721^2$. As an example far greater than that for 103, Hunter gives the subsidiary equations

$$49143\ 127346\ 631084^2 \pm 46867\ 792486\ 220437^2 = 67909\ 034288\ 072605^2 \quad \text{or} \quad 263 \times 911\ 391767\ 518393^2.$$

He has pairs $(x, y)$ for 221 of the known congruent numbers $<1000$, for example, $5829\ 043537^2 \pm 457 \times 234\ 834600^2 = 7692\ 857713^2$ or $2962\ 336463^2$ while $86236\ 037017^2 \pm 133 \times 7049\ 242860^2$ and $318\ 957135\ 928681^2 \pm 183 \times 7\ 531376\ 243820^2$ are all squares, but there remain well over a hundred for which a solution has not been constructed.

In a recent letter, Jean Lagrange observes that although 897 appears in Gérardin's tables, he believes that its character is unknown, so we have removed it from Table 7. He also reports that he has established that 113, 337, 409, and 521 are noncongruent, and hopes to be able to classify all the numbers $\equiv 1, 2,$ or $3 \pmod 8$.

Here is a summary of Table 7:

| Residue class, modulo 8 | 1 | 2 | 3 | 5 | 6 | 7 | Total |
|---|---|---|---|---|---|---|---|
| Number of known congruent numbers | 22 | 19 | 12 | 95 | 103 | 101 | 352 |
| Number of known noncongruent numbers | 68 | 76 | 87 | 0 | 0 | 0 | 231 |
| Number with status unknown | 8 | 6 | 2 | 7 | 0 | 2 | 25 |
| Total number of squarefree numbers | 98 | 101 | 101 | 102 | 103 | 103 | 608 |

Ronald Alter, The congruent number problem, *Amer. Math. Monthly* **87** (1980) 43–45.

R. Alter and T. B. Curtz, A note on congruent numbers, *Math. Comput.* **28** (1974) 303–305; *MR* **49** #2504.

R. Alter, T. B. Curtz and K. K. Kubota, Remarks and results on congruent numbers, *Congressus Numerantium VI, Proc. 3rd S. E. Conf. Combin. Graph Theory.* Comput. 1972, 27–35; *MR* **50** #2047.

L. Bastien, Nombres congruents, *Intermédiaire des Math.* **22** (1915) 231–232.

B. J. Birch, Diophantine analysis and modular functions. *Proc. Bombay Colloq. Alg. Geom.* 1968.

J. W. S. Cassels, Diophantine equations with special reference to elliptic curves, *J. London Math. Soc.* **41** (1966) 193–291.
L. E. Dickson, *History of the Theory of Numbers*, Vol. 2, Diophantine Analysis, Washington, 1920, 459–472.
A. Genocchi, Note analitiche sopra Tre Scritti *Annali di Sc. Mat. e Fis.* **6** (1855) 273–317.
A. Gérardin, Nombres congruents, *Intermédiaire des Math.* **22** (1915) 52–53.
H. J. Godwin, A note on congruent numbers, *Math. Comput.* **32** (1978) 293–295 and **33** (1979) 847; *MR* **58** #495; **80c**:10018.
Jean Lagrange, Thèse d'Etat de l'Université de Reims, 1976.
Jean Lagrange, Construction d'une table de nombres congruents, *Bull. Soc. Math. France Mém.* No. 49–50 (1977) 125–130; *MR* **58** #5498.
L. J. Mordell, *Diophantine Equations*, Academic Press, London, 1969, 71–72.
S. Roberts, Note on a problem of Fibonacci's, *Proc. London Math. Soc.* **11** (1879–80) 35–44.
N. M. Stephens, Congruence properties of congruent numbers, *Bull. London Math. Soc.* **7** (1975) 182–184; *MR* **53** #260.

**D28.** Mordell asked for the integer solutions of

$$\frac{1}{w}+\frac{1}{x}+\frac{1}{y}+\frac{1}{z}+\frac{1}{wxyz}=0.$$

L. J. Mordell, Research problem 6, *Canad. Math. Bull.* **17** (1974) 149.

# E. Sequences of Integers

Here we are mainly, but not entirely, concerned with infinite sequences; there is some overlap with sections C and A. An excellent text and source of problems is H. Halberstam and K. F. Roth, *Sequences*, Vol. I, Oxford Univ. Press, 1966. It is to be hoped that Vol. II will follow in a finite time. Other references are


P. Erdös, A. Sárközi, and E. Szemerédi, On divisibility properties of sequences of integers, in *Number Theory, Colloq. Math. Soc. János Bolyai* **2**, North-Holland, 1970, 35–49.

H. Ostmann, *Additive Zahlentheorie* I, II, Springer-Verlag, Heidelberg, 1956.

A. Stöhr, Gelöste und ungelöste Fragen über Basen der natürlichen Zahlenreihe, I, II, *J. reine angew. Math.* **194** (1955) 40–65, 111–140.

Paul Turán (ed.) *Number Theory and Analysis; a collection of papers in honor of Edmund Landau* (1877–1938), Plenum Press, New York, 1969, contains several papers, by Erdös and others, on sequences of integers.


We will denote by $\mathscr{A} = \{a_i\}$, $i = 1, 2, \ldots$ a possibly infinite strictly increasing sequence of nonnegative integers. The number of $a_i$ which do not exceed $x$ is denoted by $A(x)$. By the **density** of a sequence we will mean $\lim A(x)/x$, if it exists.

**E1.** Erdös offers $50.00 for a solution of the problem: does there exist a sequence thin enough that $A(x) < c \ln x$, but with every sufficiently large integer expressible in the form $p + a_i$, where $p$ is a prime?

The analogous problem with $r$th powers in place of primes was solved by Leo Moser; Ruzsa has solved the problem with powers of two.


L. Moser, On the additive completion of sets of integers, *Proc. Symp. Pure Math.* **8** Amer. Math. Soc. Providence, 1965, 175–180.

I. Ruzsa, On a problem of P. Erdös, *Canad. Math. Bull.* **15** (1972) 309–310.

**E2.** What is the maximum value of $A(x)$ if the least common multiple $[a_i, a_j]$ of each pair of members of the sequence is at most $x$? It is known that

$$(9x/8)^{1/2} \le \max A(x) \le (4x)^{1/2}.$$

P. Erdös, Problem, *Mat. Lapok* **2** (1951) 233.

**E3.** Is it true that the density of those integers

6, 12, 15, 18, 20, 24, 28, 30, 35, 36, 40, 42, 45, 48, 54, 56, 60, 63, 66, 70, 72, . . .

which have two divisors $d_1$, $d_2$ such that $d_1 < d_2 < 2d_1$, is one? Erdös has shown that the density exists. There is a connexion with covering congruences (F13).

P. Erdös, On the density of some sequences of integers, *Bull. Amer. Math. Soc.* **54** (1948) 685–692; *MR* **10**, 105.

**E4.** If no member of the sequence $\{a_i\}$ divides the product of $r$ other terms, Erdös shows that

$$\pi(x) + c_1 x^{2/(r+1)}(\ln x)^{-2} < \max A(x) < \pi(x) + c_2 x^{2/(r+1)}(\ln x)^{-2}$$

where $\pi(x)$ is the number of primes $\le x$. If, however, we suppose that the products of any number, not greater than $r$, of the $a_i$ are distinct, what is the max $A(x)$? For $r \ge 3$, Erdös shows

$$\max A(x) < \pi(x) + O(x^{2/3+\varepsilon}).$$

P. Erdös, On sequences of integers no one of which divides the product of two others and on some related problems, *Inst. Math. Mec. Tomsk* **2** (1938) 74–82.
P. Erdös, Extremal problems in number theory V (Hungarian), *Mat. Lapok* **17** (1966) 135–155.
P. Erdös, On some applications of graph theory to number theory, *Publ. Ramanujan Inst.* **1** (1969) 131–136.

**E5.** Let $D(x)$ be the number of numbers not greater than $x$ which are divisible by at least one $a_i$ where $a_1 < a_2 < \cdots < a_k \le n$ is a finite sequence. Is $D(x)/x < 2D(n)/n$ for all $x > n$? The number 2 cannot be reduced: for example, $n = 2a_1 - 1$, $x = 2a_1 < a_2$. In the other direction it is known that for each $\varepsilon > 0$, there is a sequence which does *not* satisfy the inequality $D(x)/x > \varepsilon D(n)/n$.

A. S. Besicovitch, On the density of certain sequences, *Math. Ann.* **110** (1934) 335–341.
P. Erdös, Note on sequences of integers no one of which is divisible by any other, *J. London Math. Soc.* **10** (1935) 126–128.

**E6.** Let $n_1 < n_2 < \cdots$ be a sequence of integers such that $n_{i+1}/n_i \to 1$ as $i \to \infty$, and the $\{n_i\}$ are distributed uniformly $(\text{mod}\, d)$ for every $d$.

[The number $N(c, d; x)$ of the $n_i \le x$ with $n_i \equiv c \pmod d$ is such that

$$N(c, d; x)/N(1, 1; x) \to 1/d \quad \text{as} \quad x \to \infty$$

for each $c$, $0 \le c \le d - 1$, and all $d$.] If $a_1 < a_2 < \cdots$ is an infinite sequence of integers for which $a_j + a_k \ne n_i$ for any $i$, $j$, $k$ then Erdös asks: is it true that the density of the $a_j$ is less than $\frac{1}{2}$?

**E7.** If $p_n$ is the $n$th prime, Erdös asks if $\sum(-1)^n n/p_n$ converges.

He also asks if, given three distinct primes and $a_1 < a_2 < a_3 < \cdots$ all the products of their powers arranged in increasing order, it is true infinitely often that $a_i$ and $a_{i+1}$ are both prime-powers. And what if we use $k$ primes or even infinitely many primes instead of three? Meyer and Tijdeman have asked a similar question for two finite sets $S$ and $T$ of primes with $a_1 < a_2 < a_3 < \cdots$ formed from $S \cup T$. Are there infinitely many $i$ for which $a_i$ is a product of powers of primes from $S$, while $a_{i+1}$ is a product of powers of primes from $T$?

**E8.** Paul Erdös and David Silverman consider $k$ integers $1 \le a_1 < a_2 < \cdots < a_k \le n$ such that no sum $a_i + a_j$ is a square. Is it true that $k < n(1 + \varepsilon)/3$, or even that $k < n/3 + O(1)$? The integers $\equiv 1 \pmod 3$ show that if this is true, it is best possible. They suggest that the same question could be asked for other sequences instead of the squares.

Erdös and Graham added to their book at the proof stage that J. P. Marsias has discovered that the sum of any two integers $\equiv 1, 5, 9, 13, 14, 17, 21, 25, 26, 29, 30 \pmod{32}$ is never a square (mod 32), so $k$ can be chosen to be at least $11n/32$. This is best possible for the modular version of the problem since J. Lagarias, A. M. Odlyzko and J. Shearer have shown that if $S \subseteq \mathbb{Z}_n$ and $S + S$ contains no square of $\mathbb{Z}_n$, then $|S| \le 11n/32$.

**E9.** K. F. Roth has conjectured that there exists an absolute constant $c$ so that for every $k$ there is an $n_0 = n_0(k)$ with the following property: For $n > n_0$, partition the integers not exceeding $n$ into $k$ classes $\{a_i^{(j)}\}$ $(1 \le j \le k)$; then the number of distinct integers not exceeding $n$ which can be written in the form $a_{i_1}^{(j)} + a_{i_2}^{(j)}$ for some $j$, is greater than $cn$.

**E10.** The well-known theorem of van der Waerden states that for every $l$ there is a number $n(h, l)$ such that if the integers not exceeding $n(h, l)$ are partitioned into $h$ classes, then at least one class contains an arithmetic progression (A.P.) containing $l + 1$ terms. More generally, given $l_0, l_1, \ldots, l_{h-1}$, there is always a class $V_i$ $(0 \le i \le h - 1)$ containing an A.P. of $l_i + 1$ terms. Denote by $W(h, l)$, or more generally $W(h; l_0, l_1, \ldots, l_{h-1})$, the least such $n(h, l)$.

Chvátal computed $W(2;2,2) = 9$, $W(2;2,3) = 18$, $W(2;2,4) = 22$, $W(2;2,5) = 32$, and $W(2;2,6) = 46$ and more recently Beeler and O'Neil give $W(2;2,7) = 58$, $W(2;2,8) = 77$, and $W(2;2,9) = 97$. The values $W(2;3,3) = 35$ and $W(2;3,4) = 55$ were also found by Chvátal and $W(2;3,5) = 73$ by Beeler and O'Neil. Stevens and Shantaram found $W(2;4,4) = 178$. Chvátal found $W(3;2,2,2) = 27$ and Brown $W(3;2,2,3) = 51$. Beeler and O'Neil also found $W(4;2,2,2,2) = 76$.

Most proofs of van der Waerden's theorem give poor estimates for $W(h,l)$. Erdös and Rado showed that $W(h,l) > (2lh^l)^{1/2}$ and Moser, Schmidt, and Berlekamp successively improved this to

$$W(h,l) > lh^{c \ln h} \quad \text{and} \quad W(h,l) > h^{l+1-c\sqrt{(l+1)\ln(l+1)}}.$$

Moser's bound has been improved for $l \geq 5$ by Abbott and Liu to

$$W(h,l) > h^{c_s(\ln h)^s}$$

where $s$ is defined by $2^s \leq l < 2^{s+1}$, and Everts has shown that $W(h,l) > lh^l/4(l+1)^2$, a result which is sometimes better than Berlekamp's.

A closely related function, with $l + 1 = k$, is the now famous $r_k(n)$, introduced long years ago by Erdös and Turán: the least $r$ such that the sequence $1 \leq a_1 < a_2 < \cdots < a_r \leq n$ of $r$ numbers not exceeding $n$ must contain a $k$-term A.P. The best bounds when $k = 3$ are due to Behrend, Roth, and Moser:

$$n \exp(-c_1\sqrt{\ln n}) < r_3(n) < c_2 n/\ln \ln n$$

and for larger $k$ Rankin showed that

$$r_k(n) > n^{1 - c_s/(\ln n)^{s/(s+1)}}$$

where $s$, much as before, is defined by $2^s < k \leq 2^{s+1}$.

A big breakthrough was Szemerédi's proof that $r_k(n) = o(n)$ for all $k$, but neither his proof, nor those of Furstenberg and of Katznelson and Ornstein (see Thouvenot) give estimates for $r_k(n)$. Erdös conjectures that

$$¿ \qquad r_k(n) = o(n(\ln n)^{-t}) \quad \text{for every } t \qquad ?$$

This would imply that for every $k$ there are $k$ primes in A.P. See A5 for a potentially remunerative conjecture of Erdös, which, if true, would imply Szemerédi's theorem.

Another closely related problem was considered by Leo Moser, who wrote the integers in base three, $n = \sum a_i 3^i$ ($a_i = 0$, 1, or 2) and considered the mapping of $n$ into lattice points $(a_1, a_2, a_3, \ldots)$ of infinite-dimensional Euclidean space. He called integers **collinear** if their images are collinear; e.g., $35 \to (2,2,0,1,0,\ldots)$, $41 \to (2,1,1,1,0,\ldots)$ and $47 \to (2,0,2,1,0,\ldots)$ are collinear. He conjectured that every sequence of integers with no three collinear has density zero. If integers are collinear, they are in A.P., but not necessarily conversely (e.g., $16 \to (1,2,1,0,0,\ldots)$, $24 \to (0,2,2,0,0,\ldots)$ and $32 \to (2,1,0,1,0,\ldots)$ are not collinear) so truth of the conjecture would imply Roth's theorem that $r_3(n) = o(n)$.

If $f_3(n)$ is the largest number of lattice points with no three in line in the $n$-dimensional cube with three points in each edge, then Moser showed that $f_3(n) > c3^n/\sqrt{n}$. It is easy to see that $f_3(n)/3^n$ tends to a limit; is it zero? Chvátal improved the constant in Moser's result to $3/\sqrt{\pi}$ and found the values $f_3(1) = 2$, $f_3(2) = 6$, $f_3(3) = 16$. It is known that $f_3(4) \geq 43$.

More generally, if the $n$-dimensional cube has $k$ points in each edge, Moser asked for an estimate of $f_k(n)$, the maximum number of lattice points with no $k$ collinear. It is a theorem of Hales and Jewett, with applications to $n$-dimensional $k$-in-a-row (tic-tac-toe), that for sufficiently large $n$, any partition of the $k^n$ lattice points into $h$ classes has a class with $k$ points in line. This implies van der Waerden's theorem on letting the point $(a_0, a_1, \ldots, a_{n-1})$, $(0 \leq a_i \leq k-1)$ correspond to the base $k$ expansion of the integer $\sum a_i k^i$. It is not known whether, for every $c$ and sufficiently large $n$, it is possible to choose $ck^n/\sqrt{n}$ points without including $k$ in line.

If you use the **greedy algorithm** to construct sequences not containing an A.P. you don't get a very dense sequence, but you do get some interesting ones. Odlyzko and Stanley construct the sequence $S(m)$ of positive integers with $a_0 = 0$, $a_1 = m$ and each subsequent $a_{n+1}$ is the least number greater than $a_n$ so that $a_0, a_1, \ldots, a_{n+1}$ does not contain a three-term A.P. For example

$S(1)$: 0, 1, 3, 4, 9, 10, 12, 13, 27, 28, 30, 31, 36, 39, 40, 81, 82, 84, 85, ...

$S(4)$: 0, 4, 5, 7, 11, 12, 16, 23, 26, 31, 33, 37, 38, 44, 49, 56, 73, 78, 80, 85, ...

If $m$ is a power of three, or twice a power of three, then the members of the sequence are fairly easy to describe in terms of their ternary expansions, but for other values the sequences behave quite erratically. Their rates of growth seem to be similar, but this has yet to be proved.

The "simplest" such sequence containing no 4-term A.P. is

0, 1, 2, 4, 5, 7, 8, 9, 14, 15, 16, 18, 25, 26, 28, 29, 30, 33, 36,
48, 49, 50, 52, 53, 55, 56, 57, 62, ...

Is there a simple description of this? How fast does it grow?

If we define the **span** of a set $S$ to be max $S$ − min $S$, what is the smallest span, sp$(k, n)$ of a set of $n$ integers containing no $k$-term A.P.? Zalman Usiskin gives the following values:

| $n =$ | 3 | 4 | 5 | 6 | 7 | 8 | 9 | 10 | 11 | ... |
|---|---|---|---|---|---|---|---|---|---|---|
| sp$(3, n) =$ | 3 | 4 | 8 | 10 | 12 | 13 | 19 | 24 | 25 | ... |
| sp$(4, n) =$ | | 4 | 5 | 7 | 8 | 9 | 12 | ... | | |

sp$(k, n)$ is the inverse of the function $r_k(n)$.

H. L. Abbott and D. Hanson, Lower bounds of certain types of van der Waerden numbers, *J. Combin. Theory* **12** (1972) 143–146.

H. L. Abbott and A. C. Liu, On partitioning integers into progression free sets, *J. Combin. Theory* **13** (1972) 432–436.

H. L. Abbott, A. C. Liu and J. Riddell, On sets of integers not containing arithmetic progressions of prescribed length, *J. Austral. Math. Soc.* **18** (1974) 188–193; *MR* **57** #12441.

Michael D. Beeler and Patrick E. O'Neil, Some new van der Waerden numbers, *Discrete Math.* **28** (1979) 135–146.

F. A. Behrend, On sets of integers which contain no three terms in arithmetical progression, *Proc. Nat. Acad. Sci. USA* **32** (1946) 331–332; *MR* **8**, 317.

E. R. Berlekamp, A construction for partitions which avoid long arithmetic progressions, *Canad. Math. Bull.* **11** (1968) 409–414.

E. R. Berlekamp, On sets of ternary vectors whose only linear dependencies involve an odd number of vectors, *Canad. Math. Bull.* **13** (1970) 363–366.

Thomas C. Brown, Some new Van der Waerden numbers, *Notices Amer. Math. Soc.* **21** (1974) A-432.

T. C. Brown, Behrend's theorem for sequences containing no $k$-element progression of a certain type, *J. Combin. Theory* Ser. A, **18** (1975) 352–356.

Ashok K. Chandra, On the solution of Moser's problem in four dimensions, *Canad. Math. Bull.* **16** (1973) 507–511.

V. Chvátal, Some unknown van der Waerden numbers, in *Combinatorial Structures and and their Applications*, Gordon and Breach, New York, 1970, 31–33.

V. Chvátal, Remarks on a problem of Moser, *Canad. Math. Bull.* **15** (1972) 19–21.

J. A. Davis, Roger C. Entringer, Ronald L. Graham and G. J. Simmons, On permutations containing no long arithmetic progressions, *Acta Arith.* **34** (1977/78) 81–90; *MR* **58** #10705.

P. Erdös, Some recent advances and current problems in number theory, in *Lectures on Modern Mathematics*, Wiley, New York, **3** (1965) 196–244.

P. Erdös and R. Rado, Combinatorial theorems on classifications of subsets of a given set, *Proc. London Math. Soc.* (3) **2** (1952) 417–439; *MR* **16**, 445.

P. Erdös and J. Spencer, *Probabilistic Methods in Combinatorics*, Academic Press, 1974, 37–39.

P. Erdös and P. Turán, On some sequences of integers, *J. London Math. Soc.* **11** (1936) 261–264.

F. Everts, PhD thesis, University of Colorado, 1977.

H. Furstenberg, Ergodic behaviour of diagonal measures and a theorem of Szemerédi on arithmetic progressions, *J. Analyse Math.* **31** (1977) 204–256; *MR* **58** #16583.

Joseph L. Gerver and Thomas L. Ramsey, Sets of integers with nonlong arithmetic progressions generated by the greedy algorithm, *Math. Comput.* **33** (1979) 1353–1359; *MR* **80k**:10053.

R. L. Graham and B. L. Rothschild, A survey of finite Ramsey Theorems, *Congressus Numerantium III*, Proc. 2nd Louisiana Conf. Combin., Graph Theory, Comput. (1971) 21–40.

R. L. Graham and B. L. Rothschild, A short proof of van der Waerden's theorem on arithmetic progressions, *Proc. Amer. Math. Soc.* **42** (1974) 385–386.

G. Hajós, Über einfache und mehrfache Bedeckung des $n$-dimensionalen Raumes mit einem Würfelgitter, *Math. Z.* **47** (1942) 427–467.

A. W. Hales and R. I. Jewett, Regularity and positional games, *Trans. Amer. Math. Soc.* **106** (1963) 222–229.

A. Y. Khinchin, *Three Pearls of Number Theory*, Graylock Press, Rochester, N.Y. 1952, 11–17.

L. Moser, On non averaging sets of integers, *Canad. J. Math.* **5** (1953) 245–252.

Leo Moser, Notes on number theory II. On a theorem of van der Waerden, *Canad. Math. Bull.* **3** (1960) 23–25; *MR* **22** #5619.

L. Moser, Problem 21, Proc. Number Theory Conf. Univ. of Colorado, Boulder, 1963, 79.

L. Moser, Problem 170, *Canad. Math. Bull.* **13** (1970) 268.
A. M. Odlyzko and R. P. Stanley, Some curious sequences constructed with the greedy algorithm, Bell. Labs. internal memo. 1978.
Carl Pomerance, Collinear subsets of lattice point sequences—an analog of Szemerédi's theorem, *J. Combin. Theory Ser. A.* **25** (1980) 140–149.
John R. Rabung, On applications of van der Waerden's theorem, *Math. Mag.* **48** (1975) 142–148.
John R. Rabung, Some progression-free partitions constructed using Folkman's method, *Canad. Math. Bull.* **22** (1979) 87–91.
R. Rado, Note on combinatorial analysis, *Proc. London Math. Soc.* **48** (1945) 122–160.
R. A. Rankin, Sets of integers containing not more than a given number of terms in arithmetical progression, *Proc. Roy. Soc. Edinburgh Sect. A* **65** (1960/61) 332–334; *MR* **26** #95.
J. Riddell, On sets of numbers containing no $l$ terms in arithmetic progressions, *Nieuw Arch. Wisk.* (3) **17** (1969) 204–209; *MR* **41** #1678.
K. F. Roth, Sur quelques ensembles d'entiers, *C.R. Acad. Sci. Paris* **234** (1952) 388–390.
K. F. Roth, On certain sets of integers, *J. London Math. Soc.* **28** (1953) 104–109; *MR* **14**, 536 (and see *ibid.* **29** (1954) 20–26; *J. Number Theory* **2** (1970) 125–142; *Period. Math. Hungar.* **2** (1972) 301–326).
R. Salem and D. C. Spencer, On sets of integers which contain no three terms in arithmetical progression, *Proc. Nat. Acad. Sci. USA* **28** (1942) 561–563; *MR* **4**, 131.
R. Salem and D. C. Spencer, On sets which do not contain a given number in arithmetical progression, *Nieuw Arch. Wisk.* (2) **23** (1950) 133–143.
H. Salié, Zur Verteilung natürlicher Zahlen auf Elementfremde Klassen, *Ber. Verh. Sächs. Akad. Wiss. Leipzig* **4** (1954) 2–26.
Wolfgang M. Schmidt, Two combinatorial theorems on arithmetic progressions, *Duke Math. J.* **29** (1962) 129–140.
G. J. Simmons and H. L. Abbott, How many 3-term arithmetic progressions can there be if there are no longer ones? *Amer. Math. Monthly* **84** (1977) 633–635; *MR* **57** #3056.
R. S. Stevens and R. Shantaram, Computer generated van der Waerden partitions, *Math. Comput.* **32** (1978) 635–636.
E. Szemerédi, On sets of integers containing no four terms in arithmetic progression, *Acta Math. Acad. Sci. Hungar.* **20** (1969) 89–104.
E. Szemerédi, On sets of integers containing no $k$ elements in arithmetic progression, *Acta Arith.* **27** (1975) 199–245.
J. P. Thouvenot, La démonstration de Furstenberg du théorème de Szemerédi sur les progressions arithmétiques, *Springer Lect. Notes in Math.* **710**, Berlin, 1979, 221–232; *MR* **81c**:10072.
B. L. van der Waerden, Beweis einer Baudet'schen Vermutung, *Nieuw Arch. voor Wisk. II* **15** (1927) 212–216.
B. L. van der Waerden, How the proof of Baudet's conjecture was found, in *Studies in Pure Mathematics*, Academic Press, London, 1971, 251–260.
E. Witt, Ein kombinatorische Satz der Elementargeometrie, *Math. Nachr.* **6** (1952) 261–262.

**E11.** Schur proved that if the integers less than $n!e$ are partitioned into $n$ classes in any way, then $x + y = z$ can be solved in integers within one class. If $s(n)$ is the least integer with this property, Abbott and Moser obtained the lower bound $s(n) > 89^{n/4 - c \ln n}$ for some $c$ and all sufficiently large $n$ and Abbott and Hanson obtained $s(n) > c(89)^{n/4}$, improving Schur's own estimate of $s(n) \geq (3^n + 1)/2$. This last result is in fact sharp for $n = 1, 2,$

and 3, but it is too low for larger values of $n$. The value $s(4) = 45$ was computed by Baumert, for example the first 44 numbers may be split into four sum-free classes

$$\{1,3,5,15,17,19,26,28,40,42,44\}, \quad \{2,7,8,18,21,24,27,33,37,38,43\},$$
$$\{4,6,13,20,22,23,25,30,32,39,41\}, \quad \{9,10,11,12,14,16,29,31,34,35,36\}.$$

More recently Fredericksen has shown that $s(5) \geq 158$ (see E12 for his example) and this improves the lower bound for all subsequent Schur numbers: $s(n) \geq c(315)^{n/5}$ $(n > 5)$.

Robert Irving has slightly improved Schur's upper bound from $\lfloor n!e \rfloor$ to $\lfloor n!(e - 1/24) \rfloor$.

H. L. Abbott, PhD thesis, University of Alberta, 1965.

H. L. Abbott and D. Hanson, A problem of Schur and its generalizations, *Acta Arith.* **20** (1972) 175–187.

H. L. Abbott and L. Moser, Sum-free sets of integers, *Acta Arith.* **11** (1966) 393–396; *MR* **34** #69.

L. D. Baumert, Sum-free sets, J. P. L. Res. Summary No 36–10, **1** (1961) 16–18.

S. L. G. Choi, The largest sum-free subsequence from a sequence of $n$ numbers, *Proc. Amer. Math. Soc.* **39** (1973) 42–44; *MR* **47** #1771.

S. L. G. Choi, J. Komlós, and E. Szemerédi, On sum-free subsequences, *Trans. Amer. Math. Soc.* **212** (1975) 307–313.

H. Fredericksen. Five sum-free sets, *Proc. 6th Ann. S.E. Conf. Graph Theory, Combin. & Comput. Congressus Numerantium XIV*, Utilitas Math. Pub. Inc 1975, 309–314.

R. W. Irving, An extension of Schur's theorem on sum-free partitions, *Acta Arith.* **25** (1973) 55–63.

J. Komlós, M. Sulyok and E. Szemerédi, Linear problems in combinatorial number theory, *Acta Math. Acad. Sci. Hungar.* **26** (1975) 113–121.

L. Mirsky, The combinatorics of arbitrary partitions, *Bull. Inst. Math. Appl.* **11** (1975) 6–9.

I. Schur, Über die Kongruenz $x^m + y^m \equiv z^m \pmod p$, *Jahresb. der Deutsche Math.-Verein.* **25** (1916) 114–117.

W. D. Wallis, A. P. Street, and J. S. Wallis, *Combinatorics: Room Squares, Sum-free Sets, Hadamard Matrices*, Springer-Verlag, Heidelberg, 1972.

Š. Znám, Generalisation of a number-theoretic result, *Mat.-Fyz. Časopis* **16** (1966) 357–361.

Š. Znám, On $k$-thin sets and $n$-extensive graphs, *Math. Časopis* **17** (1967) 297–307.

**E12.** A similar problem to Schur's was considered by Abbott and Wang. Let $t(n)$ be the least integer $m$ so that no matter how the integers from 1 to $m$ are partitioned into $n$ classes, there will always be a class containing a solution to the congruence

$$x + y \equiv z \qquad (\text{mod } (m+1))$$

Clearly $t(n) \leq s(n)$, where $s(n)$ is as in Schur's problem (E11), but for $n =$ 1, 2 or 3, we have equality, $t(1) = s(1) = 2$, $t(2) = s(2) = 5$, $t(3) = s(3) = 14$. Indeed, the only three partitions of $[1, 13]$ into three sets satisfying the sum-free condition,

$$\{1,4,10,13\} \quad \{2,3,11,12\} \quad \{5,6,8,9\}$$

(with 7 in any of the three sets) all satisfy the seemingly more restrictive congruence-free condition, modulo 14, while Baumert's example (E11) shows only one failure: $33 + 33 \equiv 21 \pmod{45}$ in the second set. In fact Baumert found 112 different ways of partitioning $[1,44]$ into four sum-free sets, and some of these are sum-free mod 45 so $t(4) = 45$. An example is

$$\{\pm 1, \pm 3, \pm 5, 15, \pm 17, \pm 19\}, \qquad \{\pm 2, \pm 7, \pm 8, \pm 18, \pm 21\},$$
$$\{\pm 4, \pm 6, \pm 13, \pm 20, \pm 22, 30\}, \qquad \{\pm 9, \pm 10, \pm 11, \pm 12, \pm 14, \pm 16\}.$$

Abbott and Wang obtained the inequality

$$f(n_1 + n_2) \geq 2f(n_1)f(n_2)$$

which holds for $f(n) = s(n) - \frac{1}{2}$ and leads to the same lower bound that Schur obtained for his problem, $t(n) \geq (3^n + 1)/2$. Indeed, they obtain evidence that $t(n) = s(n)$. Moreover, the example of Fredericksen

$$\pm\{1,4,10,16,21,23,28,34,40,43,45,48,54,60\},$$
$$\pm\{2,3,8,9,14,19,20,24,25,30,31,37,42,47,52,65,70\},$$
$$\pm\{5,11,12,13,15,29,32,33,35,36,39,53,55,56,57,59,77,79\},$$
$$\pm\{6,7,17,18,22,26,27,38,41,46,50,51,75\},$$
$$\pm\{44,49,58,61,62,63,64,67,68,69,71,72,73,74,76,78\},$$

which shows that $s(5) \geq 158$; is also sum-free mod 159 so that $t(5) \geq 158$ and $t(n) > c(315)^{n/5}$ as well.

H. L. Abbott and E. T. H. Wang, Sum-free sets of integers, *Proc. Amer. Math. Soc.* **67** (1977) 11–16; *MR* **58** #5571.

**E13.** Turán has shown that if the integers $[m, 5m + 3]$ are partitioned into two classes in any way, then in at least one of them the equation $x + y = z$ is solvable with $x \neq y$, and that this is not true for the integers $[m, 5m + 2]$. The uniqueness of the partition of $[m, 5m + 2]$ into two sum-free sets has been demonstrated by Znám.

Turán also considered the problem where $x$, $y$ are not necessarily distinct. Define $s(m, n)$ as the least integer $s$ such that however the interval $[m, m + s]$ is partitioned into $n$ classes, one of them contains a solution of $x + y = z$. His result corresponding to the first problem is $s(m, 2) = 4m$. Clearly $s(1, n) = s(n) - 1$, where $s(n)$ is as in E11, and Irving's result implies $s(m, n) \leq m\lfloor n!(e - 1/24) - 1\rfloor$. Abbott and Znám (see E11) independently noted that $s(m, n) \geq 3s(m, n) + m$ so that $s(m, n) \geq m(3^n - 1)/2$.

Abbott and Hanson call a class **strongly sum-free** if it contains no solution to either of the equations $x + y = z$ or $x + y + 1 = z$. They show that if $r(n)$ is the least $r$ such that however $[1, r]$ is partitioned into $n$ classes, one of them contains such a solution, then $r(m+n) \geq 2r(n)s(m) - r(n) - s(m) + 1$. They used this to improve the lower bound for $s(m,n)$; their method, with Fredericksen's example, now gives $s(m,n) > cm(315)^{n/5}$.

Š. Znám, Megjegyzések Turán Pál egy publikálatlan ereményéhez, *Mat. Lapok* **14** (1963) 307–310.

**E14.** Rado has considered a number of generalizations of van der Waerden's and Schur's problems. For example he shows that for any natural numbers $a$, $b$, $c$, there is a number $u$ so that however the numbers $[1,u]$ are partitioned into two classes, there is a solution of $ax + by = cz$ in at least one of the classes. He gives a value for $u$, but, as in Schur's original problem, this is not best possible. For example, for $2x + y = 5z$, the theorem gives $u = 20$, whereas it is true even for $u = 15$, though not for any smaller value of $u$: neither of the sets

$$\{1,4,5,6,9,11,14\} \quad \text{and} \quad \{2,3,7,8,10,12,13\}$$

contains a solution of $2x + y = 5z$. If we are allowed *three* sets, then 45 is the least value for $u$:

$$\{1,4,5,6,9,11,14,16,19,20,21,24,26,29,31,34,36,39,41,44\},$$
$$\{2,3,7,8,10,12,13,15,17,18,22,23,27,28,32,33,37,38,42,43\},$$
$$\{6,7,8,9,25,30,35,40\}$$

Rado called the equation $\sum a_i x_i = 0$, where the $a_i$ are nonzero integers, $n$-**fold regular** if there is a number $u(n)$, which we can assume to be minimal, such that however the interval $[1, u(n)]$ is partitioned into $n$ classes, at least one class contains a solution to the equation. He called it **regular** if it was $n$-fold regular for all $n$, and showed that an equation was regular just if $\sum a_j = 0$ for some subset of the $a_i$. For example, if $a_1 = a_2 = 1$ and $a_3 = -1$, we have Schur's original problem with $u(n) = s(n)$. Salié and Abbott considered the problem of finding lower bounds for $u(n)$; see E10 and E11 for references.

The above example, with $a_1 = 2$, $a_2 = 1$, $a_3 = -5$ is *not* regular, since, although we have seen that it is both 2-fold and 3-fold regular, it is not 4-fold regular. For, put every number $5^k l$, where $5 \nmid l$, into just one of four classes, according as $k$ is even or odd, and $l$ is $\equiv \pm 1$ or $\pm 2 \pmod 5$. It can be verified that none of these four classes contains a solution of $2x + y = 5z$.

For the equations $2x_1 + x_2 = 2x_3$ and $x_1 + x_2 + x_3 = 2x_4$, Salié, Abbott, and Abbott and Hanson obtained successively better lower bounds, culminating in $u(n) > c(12)^{n/3}$ and $c(10)^{n/3}$ respectively.

Compare problems E10–14 with C14–16.

Walter Deuber, Partitionen und lineare Gleichungssysteme, *Math. Z.* **133** (1973) 109–123.

R. Rado, Studien zur Kombinatorik, *Math. Z.* **36** (1933) 424–480.
E. R. Williams, M.Sc. thesis, Memorial Univ. 1967.

**E15.** Lenstra has remarked that the recursion $x_0 = 1$, $x_n = (1 + x_0^2 + x_1^2 + \cdots + x_{n-1}^2)/n\ (n = 1, 2, \ldots)$ [or $(n+1)x_{n+1} = x_n(x_n + n)$] yields $x_1 = 2$, $x_2 = 3$, $x_3 = 5$, $x_4 = 10$, $x_5 = 28$, $x_6 = 154$, $x_7 = 3520$, $x_8 = 1551880$, $x_9 = 267593772160, \ldots$ and $x_n$ is an integer for all $n \le 42$, but $x_{43}$ is not! Alf van der Poorten asks: what is going on? David Boyd asks whether $x_n$ is always a 2-adic integer (i.e., does not have 2 in its denominator).

The corresponding sequence with cubes in place of squares has also been examined by Boyd and van der Poorten. It appears to hold out as far as $x_{89}$, although at the time of writing, the possibility of powers of 2, 3, 5, and 7 in the denominator had not been completely ruled out.

**E16.** When he was a student L. Collatz asked if the sequence defined by $a_{n+1} = a_n/2$ ($a_n$ even), $a_{n+1} = 3a_n + 1$ ($a_n$ odd) is tree-like in structure, apart from the cycle 4, 2, 1, 4, ... (Figure 15) in the sense that, starting from any integer $a_1$, there is a value of $n$ for which $a_n = 1$. This has been verified for all integers less than $10^9$ by D. H. and Emma Lehmer and J. L. Selfridge, and by others to $7 \times 10^{11}$.

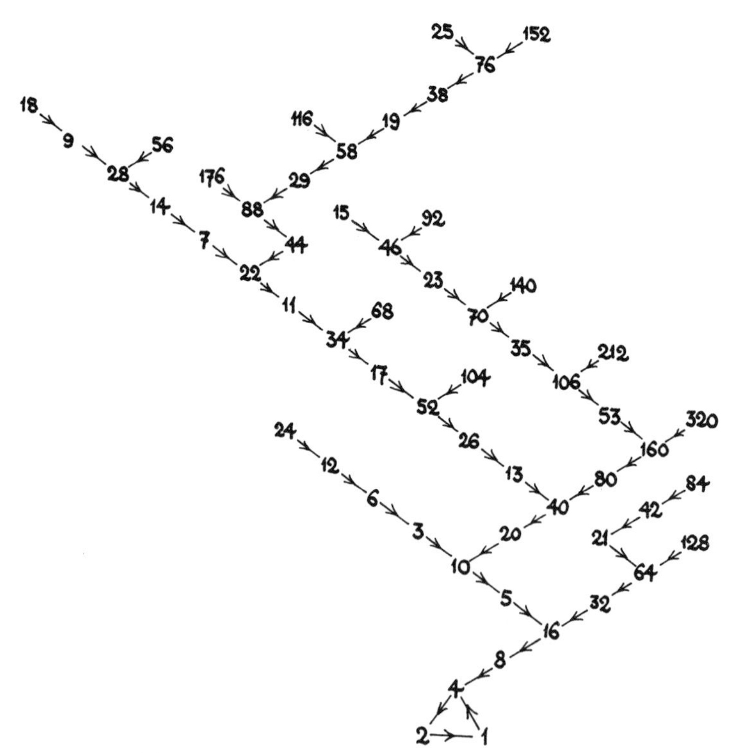


Figure 15. Is the Collatz Sequence Tree-like?

If $3a_n + 1$ is replaced by $3a_n - 1$ (or if we include negative integers), then it seems likely that any sequence concludes with one of the cycles $\{1,2\}$, $\{5,14,7,20,10\}$, or $\{17,50,25,74,37,110,55,164,82,41,122,61,182,91,272,136,68,34\}$. This is true for all numbers less than $10^8$.

David Kay defines the sequence more generally by $a_{n+1} = a_n/p$ $(p \mid a_n)$, $a_{n+1} = a_n q + r$ $(p \nmid a_n)$ and asks if there are numbers $p$, $q$, $r$ for which the problem can be settled.

Define $f(n)$ as the largest odd divisor of $3n + 1$. Zimian asked if

$$\prod_{i=1}^{m} n_i = \prod_{i=1}^{m} f(n_i)$$

holds for any (multi)set $\{n_i\}$ of integers, $n_i > 1$. Erdös found that

$$65 \times 7 \times 7 \times 11 \times 11 \times 17 \times 17 \times 13 = 49 \times 11 \times 11 \times 17 \times 17 \times 13 \times 13 \times 5.$$

Call an integer $n$ **self-contained** if $n$ divides $f^k(n)$ for some $k \geq 1$. If this happens and if the Collatz sequence $n^* = f^k(n)/n$ reaches 1, then the set

$$\{n, f(n), \ldots, f^{k-1}(n), n^*, f(n^*), \ldots, 1\}$$

is a set such as the above. A computer search for $n \leq 10^4$ yielded five self-contained integers: 31, 83, 293, 347, and 671.

These mappings are not one-one; you can't retrace the history of a sequence, since there is often no unique inverse.


Michael Beeler, William Gosper, and Rich Schroeppel, Hakmem, Memo 239, Artificial Intelligence Laboratory, M.I.T., 1972, p. 64.

Corrado Böhm and Giovanna Sontacchi, On the existence of cycles of given length in integer sequences like $x_{n+1} = x_n/2$ if $x_n$ even, and $x_{n+1} = 3x_n + 1$ otherwise, *Atti Accad. Naz. Lincei Rend. Sci. Fis. Mat. Natur.* (8) **64** (1978) 260–264.

R. E. Crandall, On the "$3x + 1$" problem, *Math. Comput.* **32** (1978) 1281–1292; *MR* **58** #494; *Zbl.* 395.10013.

C. J. Everett, Iteration of the number-theoretic function $f(2n) = n$, $f(2n + 1) = 3n + 2$, *Advances in Math.*, **25** (1977) 42–45; *MR* **56** #15552; *Zbl.* 352.10001.

Martin Gardner, Mathematical Games, A miscellany of transcendental problems, simple to state but not at all easy to solve, *Scientific Amer.* **226** #6 (Jun 1972) 114–118, esp p. 115.

E. Heppner, Eine Bemerkung zum Hasse-Syracuse-Algorithmus, *Arch. Math.* (Basel) **31** (1977/79) 317–320; *MR* **80d**:10007; *Zbl.* 377.10027.

David C. Kay, *Pi Mu Epsilon J.* **5** (1972) 338.

H. Möller, Über Hasses Verallgemeinerung der Syracuse-Algorithmus (Kakutani's Problem), *Acta Arith.* **34** (1978) 219–226; *MR* **57** #16246; *Zbl.* 329.10008.

Ray P. Steiner, A theorem on the Syracuse problem, *Congressus Numerantium XX*, Proc. 7th Conf. Numerical Math, Comput. Manitoba, 1977, 553–559; *MR* **80g**: 10003.

Riho Terras, A stopping time problem on the positive integers, *Acta Arith.* **30** (1976) 241–252; *MR* **58** #27879 (and see **35** (1979) 100–102; *MR* **80h**:10066.


**E17.** The situation is different, though no more clear, in the case of Conway's **permutation sequences**. A simple example is $a_{n+1} = 3a_n/2$ ($a_n$ even),

$a_{n+1} = \lfloor (3a_n + 1)/4 \rfloor$ ($a_n$ odd), or, perhaps more perspicuously,

$$2m \to 3m \qquad 4m - 1 \to 3m - 1 \qquad 4m + 1 \to 3m + 1$$

from which it is clear that the inverse operation works just as well. So the resulting structure consists only of disjoint cycles and doubly infinite chains. It is not known whether there is a finite or infinite number of each of these, or even whether an infinite chain exists. It is conjectured that the only cycles are $\{1\}$, $\{2, 3\}$, $\{4, 6, 9, 7, 5\}$, and $\{44, 66, 99, 74, 111, 83, 62, 93, 70, 105, 79, 59\}$. Mike Guy, with the help of TITAN, showed that any other cycles have period greater than 320. What is the status of the sequence containing the number 8?

$$\ldots, 73, 55, 41, 31, 23, 17, 13, 10, 15, 11, 8, 12, 18, 27, 20, 30, 45, 34, 51, 38, 57, 43, 32, 48, 72, \ldots$$

J. H. Conway, Unpredictable iterations, in *Proc. Number Theory Conf.*, Boulder, 1972, 49–52.

**E18.** Mahler has considered the following problem: given any real number $\alpha$, let $r_n$ be the fractional part of $\alpha(3/2)^n$. Do there exist ***Z*-numbers**, for which $0 \le r_n < \frac{1}{2}$ for all $n$? Probably not. Mahler shows that there is at most one between each pair of consecutive integers, and that, for $x$ large enough, at most $x^{0.7}$ less than $x$.

A similar question is: is there a rational number $r/s$ such that $\lfloor (r/s)^n \rfloor$ is odd for all $n$?

Littlewood once remarked that it was not known that the fractional part of $e^n$ did not tend to 0 as $n \to \infty$.

K. Mahler, An unsolved problem on the powers of 3/2, *J. Austral. Math. Soc.* **8** (1968) 313–321; *MR* **37** #2694.

**E19.** Forman and Shapiro have proved that infinitely many integers of the form $\lfloor (4/3)^n \rfloor$ and also of the form $\lfloor (3/2)^n \rfloor$ are composite. A. L. Whiteman conjectures that these two sequences also each contain infinitely many primes.

W. Forman and H. N. Shapiro, An arithmetic property of certain rational powers, *Comm. Pure Appl. Math.* **20** (1967) 561–573; *MR* **35** #2852.

**E20.** Form sequences from an alphabet $[1, n]$ of $n$ letters such that there are no immediate repetitions $\ldots aa \ldots$ and *no* alternating subsequences $\ldots a \ldots b \ldots a \ldots b \ldots$ of length greater than $d$. Denote by $N_d(n)$ the maximal length of any such sequence; then a sequence of this length is a **Davenport–Schinzel sequence**. The problem is to determine all D–S sequences, and in particular to find $N_d(n)$. We need only consider **normal**

sequences in which the first appearance of an integer of the alphabet comes after the first appearance of every smaller one.

The sequences 12131323, 121212131313232323, and 1 2 1 3 1 4 1 . . . 1 $n-1$ 1 $n-1$ $n-2$ . . . 3 2 $n$ 2 $n$ 3 $n$ . . . $n$ $n-1$ $n$ show that $N_4(3) \geq 8$, $N_8(3) \geq 20$ and $N_4(n) \geq 5n - 8$. Davenport and Schinzel showed that $N_1(n) = 1$, $N_2(n) = n$, $N_3(n) = 2n - 1$; that $N_4(n) = O(n \ln n/\ln \ln n)$, $\lim N_4(n)/n \geq 8$ and, with J. H. Conway, that $N_4(lm + 1) \geq 6lm - m - 5l + 2$, so $N_4(n) = 5n - 8$ $(4 \leq n \leq 10)$. Z. Kolba showed that $N_4(2m) \geq 11m - 13$ and Mills obtained the values of $N_4(n)$ for $n \leq 21$. For example, the sequence

$$abacadaeafafedcbgbhbhgcicigdjdjgekekgfhijklkljlilhlfl$$

is part of the proof that $N_4(12) = 53$.

Roselle and Stanton fixed $n$ rather than $d$ and obtained $N_d(2) = d$, $N_d(3) = 2\lfloor 3d/2 \rfloor - 4$ $(d > 3)$, $N_d(4) = 2\lfloor 3d/2 \rfloor + 3d - 13$ $(d > 4)$ and $N_d(5) = 4\lfloor 3d/2 \rfloor + 4d - 27$ $(d > 5)$, though Peterkin observed that this last parenthesis should be $(d > 6)$ since $N_6(5) = 34$. Roselle and Stanton also showed that normal D–S sequences of length $N_{2d+1}(5)$ are unique and that there are just two of length $N_{2d+1}(4)$ and $N_{2d}(5)$. Peterkin exhibited the 56 D–S sequences of length $N_5(6) = 29$ and showed that $N_5(n) \geq 7n - 13$ $(n > 5)$ and $N_6(n) \geq 13n - 32$ $(n > 5)$.

Rennie and Dobson gave an upper bound for $N_d(n)$ in the form

$$(nd - 3n - 2d + 7)N_d(n) \leq n(d - 3)N_d(n - 1) + 2n - d + 2 \qquad (d > 3)$$

thus generalizing the result of Roselle and Stanton for $d = 4$.

Apart from obtaining particular numerical values, the main problem is to prove or disprove $N_d(n) = O(n)$. Szemerédi proved that $N_d(n) < c_d nL(n)$ where $L$ is the least number of $e$'s for which

$$e^{e^{\cdot^{\cdot^{\cdot^{e}}}}} > n$$

Table 8. Values of $N_d(n)$.

| $d \backslash n$ | 1 | 2 | 3 | 4 | 5 | 6 | 7 | 8 | 9 | 10 | 11 | 12 | 13 | 14 | 15 | 16 | 17 | 18 | 19 | 20 | 21 |
|---|---|---|---|---|---|---|---|---|---|---|---|---|---|---|---|---|---|---|---|---|---|
| 1 | 1 | 1 | 1 | 1 | 1 | 1 | 1 | 1 | 1 | 1 | 1 | 1 | 1 | 1 | 1 | 1 | 1 | 1 | 1 | 1 | 1 |
| 2 | 1 | 2 | 3 | 4 | 5 | 6 | 7 | 8 | 9 | 10 | 11 | 12 | 13 | 14 | 15 | 16 | 17 | 18 | 19 | 20 | 21 |
| 3 | 1 | 3 | 5 | 7 | 9 | 11 | 13 | 15 | 17 | 19 | 21 | 23 | 25 | 27 | 29 | 31 | 33 | 35 | 37 | 39 | 41 |
| 4 | 1 | 4 | 8 | 12 | 17 | 22 | 27 | 32 | 37 | 42 | 47 | 53 | 58 | 64 | 69 | 75 | 81 | 86 | 92 | 98 | 104 |
| 5 | 1 | 5 | 10 | 16 | 22 | 29 | | | | | | | | | | | | | | | |
| 6 | 1 | 6 | 14 | 23 | 34 | | | | | | | | | | | | | | | | |
| 7 | 1 | 7 | 16 | 28 | | | | | | | | | | | | | | | | | |
| 8 | 1 | 8 | 20 | 35 | | | | | | | | | | | | | | | | | |
| 9 | 1 | 9 | 22 | 40 | | | | | | | | | | | | | | | | | |
| 10 | 1 | 10 | 26 | 47 | | | | | | | | | | | | | | | | | |

H. Davenport, A combinatorial problem connected with differential equations II (ed. A. Schinzel), *Acta Arith.* **17** (1971) 363–372.
H. Davenport and A. Schinzel, A combinatorial problem connected with differential equations, *Amer. J. Math.* **87** (1965) 684–694.
Annette J. Dobson and Shiela Oates Macdonald, Lower bounds for the lengths of Davenport–Schinzel sequences, *Utilitas Math.* **6** (1974) 251–257.
W. H. Mills, Some Davenport–Schinzel sequences, *Congressus Numerantium IX*, Proc. 3rd Manitoba Conf. Numerical Math. 1973, 307–313; *MR* **50** #135.
C. R. Peterkin, Some results on Davenport–Schinzel sequences, *Congressus Numerantium IX*, Proc. 3rd Manitoba Conf. Numerical Math. 1973, 337–344; *MR* **50** #136.
B. C. Rennie and A. J. Dobson, Upper bounds for the lengths of Davenport–Schinzel sequences, *Utilitas Math.* **8** (1975) 181–185.
D. P. Roselle, An algorithmic approach to Davenport–Schinzel sequences, *Utilitas Math.* **6** (1974) 91–93; *MR* **50** #9780.
D. P. Roselle and R. G. Stanton, Results on Davenport–Schinzel sequences, *Congressus Numerantium I*, Proc. Louisiana Conf. Combin. Graph Theory, Comput. Baton Rouge, 1970, 249–267.
R. G. Stanton and P. H. Dirksen, Davenport–Schinzel sequences, *Ars Combinatoria* **1** (1976) 43–51.
R. G. Stanton and R. C. Mullin, A map-theoretic approach to Davenport–Schinzel sequences, *Pacific J. Math.* **40** (1972) 167–172.
R. G. Stanton and D. P. Roselle, A result on Davenport–Schinzel sequences, Colloq. Math. Soc. Janós Bolyai **4**, *Combinatorial Theory and its Applications*, Balatonfüred, 1969, 1023–1027.
R. G. Stanton and D. P. Roselle, Some properties of Davenport–Schinzel sequences, *Acta Arith.* **17** (1970–71) 355–362.

**E21.** Thue showed that there are infinite sequences on 3 symbols which contain no two identically equal consecutive segments, and sequences on 2 symbols which contain no three identically equal consecutive segments, and many others have rediscovered these results.

If, in place of identically equal segments, we ask for no consecutive segments which are *permutations* of one another, Justin constructed a sequence on 2 symbols without five consecutive segments which are permutations of each other, and Pleasants constructed a sequence on 5 symbols without two such consecutive segments.

Dekking has solved the $(2, 4)$ and $(3, 3)$ problems, but describes the $(4, 2)$ case as "an interesting open problem." Is there a sequence on 4 symbols without consecutive segments which are permutations of each other?

S. Arshon, Démonstration de l'éxistence des suites asymétriques infinies (Russian. French summary), *Mat. Sb.* **2** (44) (1937) 769–779.
C. H. Braunholtz, Solution to problem 5030 [1962, 439], *Amer. Math. Monthly* **70** (1963) 675–676.
T. C. Brown, Is there a sequence on four symbols in which no two adjacent segments are permutations of one another? *Amer. Math. Monthly* **78** (1971) 886–888.
Richard A. Dean, A sequence without repeats on $x, x^{-1}, y, y^{-1}$, *Amer. Math. Monthly* **72** (1965) 383–385.
F. M. Dekking, On repetitions of blocks in binary sequences, *J. Combin. Theory Ser. A*, **20** (1976) 292–299.

F. M. Dekking, Strongly non-repetitive sequences and progression-free sets, *J. Combin. Theory Ser. A* **27** (1979) 181–185.
R. C. Entringer, D. E. Jackson and J. A. Schatz, On non-repetitive sequences, *J. Combin. Theory Ser. A*, **16** (1974) 159–164.
P. Erdös, Some unsolved problems, *Magyar Tud. Akad. Mat. Kutató Int. Közl.* **6** (1961) 221–254, esp. p. 240.
A. A. Evdokimov, Strongly asymmetric sequences generated by a finite number of symbols, *Dokl. Akad. Nauk SSSR* **179** (1968) 1268–1271; *Soviet Math. Dokl.* **9** (1968) 536–539.
Earl Dennet Fife, Binary sequences which contain no BBb, PhD thesis, Wesleyan Univ., Middletown, Connecticut, 1976.
D. Hawkins and W. E. Mientka, On sequences which contain no repetitions, *Math. Student* **24** (1956) 185–187; *MR* **19**, 241.
G. A. Hedlund, Remarks on the work of Axel Thue on sequences, *Nordisk Mat. Tidskr.* **15** (1967) 147–150; *MR* **37** #4454.
G. A. Hedlund and W. H. Gottschalk, A characterization of the Morse minimal set, *Proc. Amer. Math. Soc.* **16** (1964) 70–74.
J. Justin, Généralisation du théorème de van der Waerden sur les semi-groupes répétitifs, *J. Combin. Theory Ser A*, **12** (1972) 357–367.
J. Justin, Semi-groupes répétitifs, *Sém. IRIA, Log. Automat.* 1971, 101–105, 108; *Zbl.* 274.20092.
J. Justin, Characterization of the repetitive commutative semigroups, *J. Algebra* **21** (1972) 87–90; *MR* **46** #277; *Zbl.* 248.05004.
John Leech, A problem on strings of beads, *Math. Gaz.* **41** (1957) 277–278.
Marston Morse, A solution of the problem of infinite play in chess, *Bull. Amer. Math. Soc.* **44** (1938) 632.
Marston Morse and Gustav A. Hedlund, Unending chess, symbolic dynamics and a problem in semigroups, *Duke Math. J.* **11** (1944) 1–7; *MR* **5** (1944) 202.
P. A. B. Pleasants, Non-repetitive sequences, *Proc. Cambridge Philos. Soc.* **68** (1970) 267–274.
Helmut Prodinger and Friedrich J. Urbanek, Infinite 0-1 sequences without long adjacent identical blocks, *Discrete Math.* **28** (1979) 277–289.
H. E. Robbins, On a class of recurrent sequences, *Bull. Amer. Math. Soc.* **43** (1937) 413–417.
A. Thue, Über unendliche Zeichenreihen, *Norse Vid. Selsk. Skr. I Mat.-Nat. Kl. Christiania* (1906), No. 7, 1–22.
A. Thue, Über die gegenseitige Lage gleicher Teile gewisser Zeichenreihen, *ibid.* (1912), No. 1, 1–67.

**E22.** Hansraj Gupta asked, for $n \geq 2$, to find the least positive integer $m = m(n)$ for which a cycle $a_1, a_2, \ldots, a_m$ of positive integers, each $\leq n$, exists such that any given permutation of the first $n$ natural numbers appears as a subsequence (not necessarily consecutive) of

$$a_j, a_{j+1}, \ldots, a_m, a_1, a_2, \ldots, a_{j-1}$$

for at least one $j$, $1 \leq j \leq m$. For example, for $n = 5$, such a cycle is 1, 2, 3, 4, 5, 4, 3, 2, 1, 5, 4, 5, so that $m(5) \leq 12$. He conjectures that $m(n) \leq \lfloor n^2/2 \rfloor$.

Motzkin and Straus used the **ruler function** (exponent of the highest power of 2 which divides $k$), e.g., $n = 5$, $1 \leq k \leq 31$,

1, 2, 1, 3, 1, 2, 1, 4, 1, 2, 1, 3, 1, 2, 1, 5, 1, 2, 1, 3, 1, 2, 1, 4, 1, 2, 1, 3, 1, 2, 1,

but this doesn't make use of the cyclic options and gives only $m(n) \leq 2^n - 1$.

**E23.** If $S$ is the union of $n$ arithmetic progressions of integers, each with common difference $\geq k$, where $k \leq n$, Crittenden and Vanden Eynden conjecture that $S$ contains all positive integers whenever it contains those $\leq k2^{n-k+1}$. If this is true, it's best possible. They have proved it for $k = 1$.

R. B. Crittenden and C. L. Vanden Eynden, Any $n$ arithmetic progressions covering the first $2^n$ integers covers all integers, *Proc. Amer. Math. Soc.* **24** (1970) 475–481.
R. B. Crittenden and C. L. Vanden Eynden, The union of arithmetic progressions with differences not less than $k$, *Amer. Math. Monthly* **79** (1972) 630.

**E24.** Erdös and Straus call a sequence of positive integers $\{a_n\}$ an **irrationality sequence** if $\sum 1/a_n b_n$ is irrational for all integer sequences $\{b_n\}$. What are the irrationality sequences? Find some interesting ones. If $\limsup(\log_2 \ln a_n)/n > 1$, where the log is to base 2, then $\{a_n\}$ is an irrationality sequence. $\{n!\}$ is not an irrationality sequence, because $\sum 1/n!(n+2) = \frac{1}{2}$. Erdös has shown that $\{2^{2^n}\}$ is an irrationality sequence. Is the sequence 2, 3, 7, 43, 1807, ... where $a_{n+1} = a_n^2 - a_n + 1$, an irrationality sequence?

P. Erdös, Some problems and results on the irrationality of the sum of infinite series, *J. Math. Sci.* **10** (1975) 1–7.

**E25.** David Silverman defined $f(1) = 1$ and $f(n)$ as the number of occurrences of $n$ in a nondecreasing sequence of integers:

| $n$ | 1 | 2 | 3 | 4 | 5 | 6 | 7 | 8 | 9 | 10 | 11 | 12 | 13 | 14 | 15 | 16 | 17 | 18 | 19 | 20 |
|---|---|---|---|---|---|---|---|---|---|---|---|---|---|---|---|---|---|---|---|---|
| $f(n)$ | 1 | 2 | 2 | 3 | 3 | 4 | 4 | 4 | 5 | 5 | 5 | 6 | 6 | 6 | 6 | 7 | 7 | 7 | 7 | 8 |

| $n$ | 21 | 22 | 23 | 24 | 25 | 26 | 27 | 28 | 29 | 30 | 31 | 32 | 33 | 34 | 35 | 36 | 37 | 38 | 39 | 40 |
|---|---|---|---|---|---|---|---|---|---|---|---|---|---|---|---|---|---|---|---|---|
| $f(n)$ | 8 | 8 | 8 | 9 | 9 | 9 | 9 | 9 | 10 | 10 | 10 | 10 | 10 | 11 | 11 | 11 | 11 | 11 | 12 | 12 |

Is there a closed expression for $f(n)$? Is $\tau^{2-\tau}n^{\tau-1}$, where $\tau = (1+\sqrt{5})/2$, a good asymptotic expression?

**E26.** Richard Epstein's Put-or-Take-a-Square game is played with one heap of chips. Two players alternately add or take away the largest perfect square number of chips that is in the heap. That is, the two players alternately name nonnegative integers $a_n$, where $a_{n+1} = a_n \pm \lfloor\sqrt{a_n}\rfloor^2$, the winner being the first to name zero. This is a loopy game and many numbers lead to a draw. For example, from 2 the next player will not take 1, allowing his opponent to win, so he goes to 3. Now to add 1 is a bad move, so the opponent goes back to 2. Similarly 6 leads to a draw with best play: 6, 10, 19!, 35, 60, 109!, 209!, 13!, 22!, 6, ... where ! means a good move, not factorial! For example, after 209, 405 is bad since the next player can go to 5 which is a $\mathscr{P}$**-position** (previous player winning). Similarly, from 60, it is bad to go to 11, an $\mathscr{N}$**-position**, one in which the next player wins (by going to 20).

Do the $\mathscr{P}$-positions

0, 5, 20, 29, 45, 80, 101, 116, 135, 145, 165, 173, 236,
257, 397, 404, 445, 477, 540, 565, 580, 585, 629, 666,
836, 845, 885, 909, 944, 949, 954, 975, 1125, 1177, ...

or the $\mathscr{N}$-positions

1, 4, 9, 11, 14, 16, 21, 25, 30, 36, 41, 44, 49, 52, 54, 64, 69, 71, 81,
84, 86, 92, 100, 105, 120, 121, 126, 136, 141, 144, 149, 164, 169,
174, 189, 196, 201, 208, 216, 225, 230, 245, 252, 254, 256, 261, ...

have positive density?

E. R. Berlekamp, J. H. Conway, and R. K. Guy, *Winning Ways*, Academic Press, London, 1981, Chap. 15.

**E27.** In his master's thesis, Roger Eggleton discussed **max sequences**, in which a given finite sequence $a_0, a_1, \ldots, a_n$ is successively extended by defining $a_{n+1} = \max_i(a_i + a_{n-i})$. One of the main results is that the first differences are ultimately periodic. For example, starting from 1, 4, 3, 2 we get 7, 8, 11, 12, 15, 16, ... with differences 3, $-1$, $-1$, 5, 1, 3, 1, 3, 1, .... What happens to **mex sequences**, where the **mex** of a set of integers is the minimum excluded number, or least nonnegative integer which does not appear in the set? Now the sequence 1, 4, 3, 2 continues

0, 0, 0, 0, 0, 5, 1, 1, 1, 1, 1, 6, 2, 2, 0, 0, 0, 0, 0, 5, 1, 1, 1, 1, 1, 6, ....

Are such sequences ultimately periodic?

Motivation for this problem comes from the analysis of *octal games* (see the Game Theory chapter in a forthcoming volume in this series) using the Sprague-Grundy theory, where ordinary addition is replaced by *nim-addition*. Now 1, 4, 3, 2 leads to

0, 0, 0, 0, 0, 5, 1, 4, 1, 1, 1, 3, 6, 6, 6, 3, 0, 2, 2, 2, 7, 2, 4, 1, ....

The behavior of such sequences remains a considerable mystery, clarification of which would lead to results about nim-like games.

E. R. Berlekamp, J. H. Conway and R. K. Guy, *Winning Ways*, Academic Press, London, 1981, Chapter 4.
R. B. Eggleton, Generalized integers, M.A. Thesis, Univ. of Melbourne, 1969.

**E28.** Call an infinite sequence $1 \le a_1 < a_2 < \cdots$ an $A$-**sequence** if no $a_i$ is the sum of distinct members of the sequence other than $a_i$. Erdös proved that for every $A$-sequence, $\sum 1/a_i < 103$, and Levine and O'Sullivan improved this to 4. They also gave an $A$-sequence whose sum of reciprocals is $>2.035$, and conjecture that this is near the right answer for the maximum.

If $1 \leq a_1 < a_2 < \cdots$ is a $B_2$-**sequence** (compare C9) i.e., a sequence where all the sums $a_i + a_j$ are different, what is the maximum of $\sum 1/a_i$? There are two problems, according as $i = j$ is permitted or not, but Erdös was unable to solve either of them.

The most obvious $B_2$-sequence is that obtained by the greedy algorithm (compare E10; each term is the least integer greater than earlier terms which does not violate the distinctness of sums condition; $i = j$ is permitted):

$$1, 2, 4, 8, 13, 21, 31, 45, 66, 81, 97, 123, 148, 182, \ldots$$

Mian and Chowla used this to show the existence of a $B_2$-sequence with $a_k \ll k^3$. If $M$ is the maximum of $\sum 1/a_i$ over all $B_2$-sequences and $S^*$ is the sum of the reciprocals of the Mian–Chowla sequence, then $M \geq S^* > 2.156$. But Levine observes that if $t_n = n(n+1)/2$, then $M \leq \sum 1/(t_n + 1) < 2.374$, and would like to see a proof or disproof of

$$¿ \quad M = S^*. \quad ?$$

Eugene Levine, An extremal result for sum-free sequences, *J. Number Theory*, **12** (1980) 251–257.

Eugene Levine and Joseph O'Sullivan, An upper estimate for the reciprocal sum of a sum-free sequence, *Acta Arith.* **34** (1977) 9–24; *MR* **57** #5900; *Zbl.* 335.10053.

Abdul Majid Mian and S. D. Chowla, On the $B_2$ sequences of Sidon, *Proc. Nat. Acad. Sci. India Sect. A* **14** (1944) 3–4; *MR* **7**–243.

J. O'Sullivan, On reciprocal sums of sum-free sequences, PhD thesis, Adelphi Univ. 1973.

**E29.** Partition the integers into two classes. Is it true that there is always a sequence $\{a_i\}$ so that all the sums $\sum \varepsilon_i a_i$ and all the products $\prod a_i^{\varepsilon_i}$ ($\varepsilon_i = 0$ or 1, all but a finite number zero) are in the same class? Hindman answered this question of Erdös negatively.

Is there a sequence $a_1 < a_2 < \cdots$ so that all the sums $a_i + a_j$ and products $a_i a_j$ are in the same class? Graham proved that if we divide the integers $[1, 252]$ into two classes, there are four distinct numbers $x$, $y$, $x + y$ and $xy$ all in the same class. Moreover, 252 is best possible. Hindman proved that if we divide the integers $[2, 990]$ into two classes, then one class always contains four distinct numbers $x$, $y$, $x + y$, $xy$. No corresponding result is known for the integers $\geq 3$.

Hindman also proved that if we divide the integers into two classes, there is always a sequence $\{a_i\}$ so that all the sums $a_i + a_j$ ($i = j$ permitted) are in the same class. On the other hand he found a decomposition into three classes so that no such infinite sequence exists.

J. Baumgartner, A short proof of Hindman's theorem, *J. Combin. Theory Ser. A* **17** (1974) 384–386.

Neil Hindman, Finite sums with sequences within cells of a partition of $n$, *J. Combin. Theory Ser. A* **17** (1974) 1–11.

Neil Hindman, Partitions and sums and products of integers, *Trans. Amer. Math. Soc.* **247** (1979) 227–245; *MR* **80b**:10022.
Neil Hindman, Partitions and sums and products—two counterexamples, *J. Combin. Theory Ser. A* **29** (1980) 113–120.

## E30. MacMahon's "prime numbers of measurement,"

1, 2, 4, 5, 8, 10, 14, 15, 16, 21, 22, 25, 26, 28, 33, 34, 35, 36, 38, 40, 42, . . .

are generated by excluding all the sums of two or more consecutive earlier members of the sequence.

If $m_n$ is the $n$th member of the sequence, and $M_n$ the sum of the first $n$ members, then George Andrews conjectures that $m_n \sim n \ln n/\ln\ln n$ and $M_n \sim n^2(\ln n)/\ln(\ln n)^2$, and poses the following, presumably easier, problems: prove $\lim n^{-\Delta}m_n = 0$ for some $\Delta < 2$; prove $\lim m_n/n = \infty$; prove $m_n < p_n$ for every $n$, where $p_n$ is the $n$th prime.

Jeff Lagarias suggests excluding only the sums of *two* or *three* consecutive earlier members, and asks if the resulting sequence

1, 2, 4, 5, 8, 10, 12, 14, 15, 16, 19, 20, 21, 24, 25, 27, 28, 32, 33, 34,
37, 38, 40, 42, 43, 44, 46, 47, 48, 51, 53, 54, 56, 57, 58, 59, 61, . . .

has density $\frac{3}{5}$.

G. E. Andrews, MacMahon's prime numbers of measurement, *Amer. Math. Monthly* **82** (1975) 922–923.
R. L. Graham, Problem #1910, *Amer. Math. Monthly* **73** (1966) 775; solution, **75** (1968) 80–81.
Jeff Lagarias, Problem 17, W. Coast Number Theory Conf., Asilomar, 1975.
P. A. MacMahon, The prime numbers of measurement on a scale, *Proc. Cambridge Philos. Soc.* **21** (1923) 651–654.
Štefan Porubský, On MacMahon's segmented numbers and related sequences, *Nieuw Arch. Wisk.* (3) **25** (1977) 403–408; *MR* **58** #5575.
N. J. A. Sloane, *A Handbook of Integer Sequences*, Academic Press, New York, 1973; sequences 363, 416, 1044.

## E31. Doug Hofstadter has defined three intriguing sequences.

(a) $a_1 = a_2 = 1$ and $a_n = a_{n-a_{n-1}} + a_{n-a_{n-2}}$ for $n \geq 3$. What is the general behavior of this sequence?

1, 1, 2, 3, 3, 4, 5, 5, 6, 6, 6, 8, 8, 8, 10, 9, 10, 11, 11, 12,
12, 12, 12, 16, 14, 14, 16, 16, 16, 16, 20, 17, 17, . . .

Are there infinitely many integers 7, 13, 15, 18, . . . that get missed out?.

(b) $b_1 = 1$, $b_2 = 2$, and for $n \geq 3$, $b_n$ is the least integer greater than $b_{n-1}$ which can be expressed as the sum of two or more *consecutive* terms of the sequence, so it goes

1, 2, 3, 5, 6, 8, 10, 11, 14, 16, 17, 18, 19, 21, 22, 24, 25,
29, 30, 32, 33, 34, 35, 37, 40, 41, 43, 45, 46, 47, . . .

This is a sort of dual of MacMahon's prime numbers of measurement (E30). How does the sequence grow?

(c) $c_1 = 2$, $c_2 = 3$, and when $c_1, \ldots, c_n$ are defined, form all possible expressions $c_i c_j - 1$ $(1 \le i < j \le n)$ and append them to the sequence:

$$2, 3, 5, 9, 14, 17, 26, 27, 33, 41, 44, 50, 51, 53, 69, 77,$$
$$80, 81, 84, 87, 98, 99, 101, 105, 122, 125, 129, \ldots$$

Does the result include almost all of the integers?

P. Erdös and R. L. Graham, *Old and New Problems and Results in Combinatorial Number Theory*, Monographie de L'Enseignement Mathématique No. 28, Genève, 1980, pp. 83–84.

**E32.** An old problem of Dickson is still unsolved. Given a set of $k$ integers, $a_1 < a_2 < \cdots < a_k$, define $a_{n+1}$ for $n \ge k$ as the least integer greater than $a_n$ which is *not* of the form $a_i + a_j$, $i, j \le n$. Except for the prescribed section at the beginning, these are $B_2$-sequences formed by the greedy algorithm (compare C9, E10, E28).

Is the sequence of differences $a_{n+1} - a_n$ eventually periodic?

Such sequences may take a long time before the periodicity appears. For example, even for $k = 2$, if we take $a_1 = 1$, $a_2 = 6$, the sequence is

$$1, 6, 8, 10, 13, 15, 17, 22, 24, 29, 31, 33, 36, 38,$$
$$40, 45, 47, 52, 54, 56, 59, 61, 63, 68, \ldots$$

and one can be forgiven for not immediately recognizing the pattern. Try starting with the set $\{1, 4, 9, 16, 25\}$.

L. E. Dickson, The converse of Waring's problem, *Bull. Amer. Math. Soc.* **40** (1934) 711–714.

**E33.** Erdös and Graham say that a sequence $\{a_i\}$ has a **monotone** A.P. of length $k$ if there are subscripts $i_1 < i_2 < \cdots < i_k$ such that the subsequence $a_{i_j}$, $1 \le j \le k$ is either an increasing or a decreasing A.P. If $M(n)$ is the number of permutations of $[1, n]$ which have no monotone 3-term A.P., then Davis et al have shown that

$$M(n) \ge 2^{n-1} \qquad M(2n-1) \le (n!)^2 \qquad M(2n+1) \le (n+1)(n!)^2.$$

They ask if $M(n)^{1/n}$ is bounded.

Davis et al have also shown that any permutation of (all) the positive integers must contain an increasing 3-term A.P., but there are permutations with no monotone 5-term A.P. It is not known whether a monotone 4-term A.P. must always occur.

If the positive integers are arranged as a doubly-infinite sequence then a monotone 3-term A.P. must still occur, but it's possible to prevent the occurrence of 4-term ones.

If *all* the integers are to be permuted then Tom Odda has shown that no 7-term A.P. need occur in the singly-infinite case, but little else is known.

J. A. Davis, R. C. Entringer, R. L. Graham and G. J. Simmons, On permutations containing no long arithmetic progressions, *Acta Arith.* **34** (1977) 81–90; *MR* **58** #10705.

Tom Odda, Solution to Problem E2440, *Amer. Math. Monthly* **82** (1975) 74.

# F. None of the Above

The first few problems in this miscellaneous section are about **lattice points**, whose Euclidean coordinates are integers. Most of them are two-dimensional problems, but some can be formulated in higher dimensions as well. Some interesting books are

J. W. S. Cassels, *Introduction to the Geometry of Numbers*, Springer-Verlag, N.Y. 1972.
L. Fejes Tóth, *Lagerungen in der Ebene, auf der Kugel und in Raum*, Springer-Verlag, Berlin, 1953.
J. Hammer, *Unsolved Problems Concerning Lattice Points*, Pitman, 1977.
O. H. Keller, *Geometrie der Zahlen*, Enzyclopedia der Math. Wissenschaften, Vol. 12, B. G. Teubner, Leipzig, 1954.
C. G. Lekkerkerker, *Geometry of Numbers*, Bibliotheca Mathematica, Vol. 8, Walters-Noordhoff, Groningen; North-Holland, Amsterdam, 1969.
C. A. Rogers, *Packing and Covering*, Cambridge Univ. Press, 1964.

**F1.** A very difficult unsolved problem is **Gauss's problem**. How many lattice points are there inside the circle with centre at the origin and radius $r$? If the answer is $\pi r^2 + h(r)$, then Hardy and Landau showed that $h(r)$ is *not* $o(r^{1/2}(\ln r)^{1/4})$. It is conjectured that $h(r) = O(r^{1/2+\varepsilon})$. The best that is known is $h(r) = O(r^{24/37+\varepsilon})$. One can ask analogous questions in three dimensions for the sphere and for the regular tetrahedron.

Chen Jing-run, The lattice points in a circle, *Sci. Sinica* **12** (1963) 633–649; *MR* **27** #4799.

**F2.** What is the largest number $k$, of lattice points $(x, y)$, $1 \le x, y \le n$, which can be chosen so that their $\binom{k}{2}$ mutual distances are all distinct? It is easy to see that $k \le n$. This bound can be attained for $n \le 7$, but not for any larger value of $n$. Erdös and Guy showed that

$$n^{2/3-\varepsilon} < k < cn/(\ln n)^{1/4}$$

and they conjecture that

$$¿ \qquad k < cn^{2/3}(\ln n)^{1/6} \qquad ?$$

One can also ask for "saturated" configurations, containing a *minimum* number of points which determine distinct distances, but such that *no* lattice point may be added without duplicating a distance. Erdös observes that this needs at least $n^{2/3-\varepsilon}$ lattice points. In one dimension he cannot improve on $O(n^{1/3})$ and suspects that $O(n^{1/2+\varepsilon})$ is best possible.

P. Erdös and R. K. Guy, Distinct distances between lattice points, *Elem. Math.* **25** (1970) 121–123.

**F3.** Erdös and Purdy ask how many of the $n^2$ lattice points $(x, y)$, $1 \le x, y \le n$ can you choose with no four of them on a circle. It is easy to show $n^{2/3-\varepsilon}$, but more should be possible.

What is the smallest $t$ so that you can choose $t$ of the points so that the $\binom{t}{2}$ lines that they determine contain all the $n^2$ lattice points? Is $t = o(n)$? It is not hard to show $t > cn^{2/3}$.

**F4. The no-three-in-line problem.** Can $2n$ lattice points $(x, y)$ $(1 \le x, y \le n)$ be selected with no three in a straight line? This has been achieved for $n \le 26$. Guy and Kelly make four conjectures.

(1) There are no configurations with the symmetry of a rectangle which do not have the full symmetry of the square.
(2) The only configurations having the symmetry of the square are those for $n = 2$, 4, and 10. (See Figure 16. The $n = 10$ configuration was first found by Acland-Hood.)

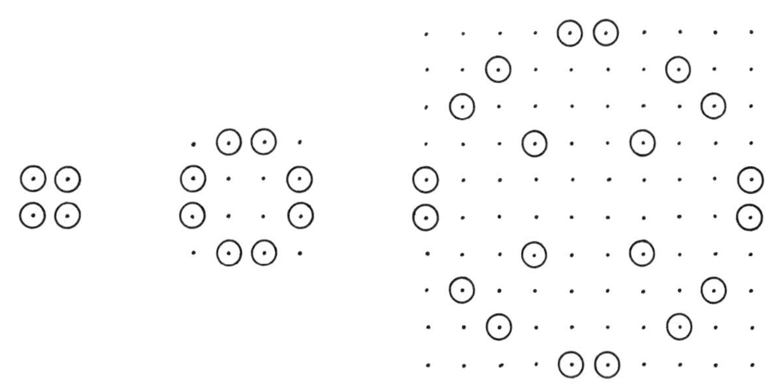

Figure 16. $2n$ Lattice Points, No Three in Line, $n = 2, 4, 10$.

(3) For large enough $n$, the answer to the initial question is "no," i.e., there are only finitely many solutions to the problem.
(4) For large $n$ we may select at most $(c + \varepsilon)n$ lattice points with no three in line, where $3c^3 = 2\pi^2$, i.e., $c \approx 1.85\ldots$ .

In the opposite direction, Erdös showed that if $n$ is prime, it is possible to choose $n$ points with no three in line and Hall et al have shown that for $n$ large, $(\frac{3}{2} - \varepsilon)n$ such points can be found.

The no-three-in-line problem is a discrete analog of a thirty-year-old problem of Heilbronn. Place $n$ ($\geq 3$) points in a disc of unit area (or unit square, or triangle of unit area) so as to maximize the area of the smallest triangle formed by three of the points. If we denote this maximum area by $\Delta(n)$, then Heilbronn's original conjecture is that $\Delta(n) < c/n^2$, and it remains an open question. Erdös showed that $\Delta(n) > 1/2n^2$, so if the conjecture is true, it is best possible. K. F. Roth showed that $\Delta(n) \ll 1/n(\ln \ln n)^{1/2}$ and Schmidt improved this to $\Delta(n) \ll 1/n(\ln n)^{1/2}$. Roth subsequently improved this to $\Delta(n) \ll 1/n^{\mu-\varepsilon}$, first with $\mu = 2 - 2/\sqrt{5} > 1.1055$ and later with $\mu = (17 - \sqrt{65})/8 > 1.1172$.

In a forthcoming paper, Komlós, Pintz and Szemerédi improve the lower bound to $(\ln n)/n^2$ thus disproving Heilbronn's conjecture. They also improve the upper bound.


Acland-Hood, *Bull. Malayan Math. Soc.* **0** (1952–53) E 11–12.

Michael A. Adena, Derek A. Holton and Patrick A. Kelly, Some thoughts on the no-three-in-line problem, Proc. 2nd. Austral. Conf. Combin. Math., *Springer Lecture Notes* **403** (1974) 6–17; *MR* **50** #1890.

David Brent Anderson, Update on the no-three-in-line problem, *J. Combin. Theory Ser A* **27** (1979) 365–366.

W. W. Rouse Ball and H. S. M. Coxeter, *Mathematical Recreations and Essays*, 12th ed. Univ. of Toronto Press, 1974, p. 189.

D. Craggs and R. Hughes-Jones, On the no-three-in-line problem, *J. Combin. Theory Ser. A* **20** (1976) 363–364; *MR* **53** #10590.

H. E. Dudeney, *The Tribune*, 1906:11:07.

H. E. Dudeney, *Amusements in Mathematics*, Nelson, London, 1917, 94, 222.

Martin Gardner, Mathematical Games: Challenging chess tasks for puzzle buffs and answers to the recreational puzzles, *Sci. Amer.* **226** #5 (May 1972) 112–117, esp. pp. 113–114.

Martin Gardner, Mathematical Games: Combinatorial problems, some old, some new and all newly attacked by computer, *Sci. Amer.* **235** #4 (Oct 1976) 131–137, esp. pp. 133–134 also **236** #3 (Mar 1977) 139–140.

Michael Goldberg, Maximizing the smallest triangle made by $N$ points in a square, *Math. Mag.* **45** (1972) 135–144.

Richard K. Guy, *Bull. Malayan Math. Soc.* **0** (1952–53) E 22.

Richard K. Guy, Unsolved combinatorial problems, in D. J. A. Welsh, *Combinatorial Mathematics and Its Applications*, Academic Press, London, 1971, p. 124.

Richard K. Guy and Patrick A. Kelly, The no-three-in-line problem, *Canad. Math. Bull.* **11** (1968) 527–531.

R. R. Hall, T. H. Jackson, A. Sudbery, and K. Wild, Some advances in the no-three-in-line problem, *J. Combin. Theory Ser. A* **18** (1975) 336–341.

P. A. Kelly, The use of the computer in game theory, M. Sc. thesis, Univ. of Calgary, 1967.

Torleiv Kløve, On the no-three-in-line problem II, *J. Combin. Theory Ser. A* **24** (1978) 126–127; *MR* **57** #2962; *Zbl.* 393.05004.
Torleiv Kløve, On the no-three-in-line problem III, *J. Combin. Theory Ser. A* **26** (1979) 82–83; *Zbl.* 393.05005.
Carl Pomerance, Collinear subsets of lattice point sequences—an analog of Szemerédi's theorem, *J. Combin. Theory Ser. A* **28** (1980) 140–149.
K. F. Roth, On a problem of Heilbronn, *J. London Math. Soc.* **25** (1951) 198–204, esp. Appendix p. 204; II, III *Proc. London Math. Soc.* **25** (1972) 193–212, 543–549.
K. F. Roth, Developments in Heilbronn's triangle problem, *Advances in Math.* **22** (1976) 364–385; *MR* **55** #2771.
Wolfgang M. Schmidt, On a problem of Heilbronn, *J. London Math. Soc.* **4** (1971/72) 545–550.

**F5.** The **quadratic residues** of a prime $p$ are the nonzero numbers $r$ for which the congruence $r \equiv x^2 \pmod p$ has solutions. They are $\frac{1}{2}(p-1)$ in number, and symmetrically distributed if $p$ is of the form $4k+1$. If $p = 4k-1$, there are more quadratic residues in the interval $[1, 2k-1]$ than in $[2k, 4k-2]$, but all known proofs use Dirichlet's class-number formula. Is there an elementary proof?

For the first few values of $d$ it is easy to remember which primes have $d$ as a quadratic residue:

| | | | |
|---|---|---|---|
| $d=-1$ | $p=4k+1$ | | |
| $d=-2$ | $p=8k+1, 3$ | $d=2$ | $p=8k\pm 1$ |
| $d=-3$ | $p=6k+1$ | $d=3$ | $p=12k\pm 1$ |
| $d=-5$ | $p=20k+1, 3, 7, 9$ | $d=5$ | $p=10k\pm 1$ |
| $d=-6$ | $p=24k+1, 5, 7, 11$ | $d=6$ | $p=24k\pm 1, 5.$ |

However, it's accidental that in these small cases the residues are just those in the first half or the end quarters of the period, according to the sign of $d$. The **Legendre symbol**, $(\frac{a}{p})$, is often used to indicate the quadratic character of a number $a$, $(a,p)=1$, relative to the prime $p$. Its value is $\pm 1$ according as $a$ is, or is not, a quadratic residue of $p$. For example, $(\frac{-1}{p}) = \pm 1$ according as $p = 4k \pm 1$.

A useful generalization of the Legendre symbol is the **Jacobi symbol** $(\frac{a}{b})$ which is defined for $(a,b)=1$ and $b$ any odd number, by the product

$$\prod \left(\frac{a}{p_i}\right)$$

of Legendre symbols, where $b = \prod p_i$ is the prime factorization of $b$, *with repetitions counted appropriately*. The important properties of these symbols include the fact that $(\frac{a}{p}) = (\frac{c}{p})$ if $a \equiv c \pmod p$; and Gauss's famous **quadratic reciprocity law**, that for odd primes $p$ and $q$, $(\frac{p}{q}) = (\frac{q}{p})$ unless $p$ and $q$ are both $\equiv -1 \pmod 4$, in which case $(\frac{p}{q}) = -(\frac{q}{p})$. These can be used for making quick verifications of the quadratic character of quite large numbers. For example, 224 is a residue of 325, because

$$\left(\frac{224}{325}\right)=\left(\frac{224}{5\times 5\times 13}\right)=\left(\frac{224}{5}\right)\left(\frac{224}{5}\right)\left(\frac{224}{13}\right)$$
$$=\left(\frac{224}{13}\right)=\left(\frac{3}{13}\right)=\left(\frac{13}{3}\right)=\left(\frac{1}{3}\right)=1;$$

in fact $224 \equiv 43^2 \pmod{325}$.

If $R$ (respectively $N$) is the maximum number of consecutive quadratic residues (respectively nonresidues) modulo an odd prime $p$, then A. Brauer showed that for $p \equiv 3 \pmod 4$, $R = N < \sqrt{p}$. On the other hand, if $p = 13$, $N = 4 > \sqrt{13}$, since 5, 6, 7, 8 are all nonresidues of 13. Schur conjectured that $N < \sqrt{p}$ if $p$ is large enough. Hudson proved Schur's conjecture; moreover, he believes that $p = 13$ is the only exception.

A. Brauer, Über die Verteilung der Potenzreste, *Math. Z.* **35** (1932) 39–50; *Zbl.* **3**, 339.
H. Davenport, *The Higher Arithmetic*, Hutchinson's Univ. Library, 1952, 74–78.
Richard H. Hudson, On sequences of quadratic nonresidues, *J. Number Theory* **3** (1971) 178–181; *MR* **43** #150.
Richard H. Hudson, On a conjecture of Issai Schur, *J. reine angew. Math.* **289** (1977) 215–220; *MR* **58** #16481.

**F6.** What patterns of quadratic residues are sure to occur? It is easy to see that a pair of neighboring ones always do, since at least one of 2, 5, and 10 is a residue, so (1, 2), (4, 5), or (9, 10) is such a pair. In the same way at least one of (1, 3), (2, 4), or (4, 6) is a pair of residues differing by 2; (1, 4) is a pair differing by 3; (1, 5), (4, 8), (6, 10), or (12, 16) is a pair differing by 4; and so on.

Suppose that each of $r$, $r + a$, $r + b$ is a quadratic residue modulo $p$. Emma Lehmer asks: for which pairs $(a, b)$ will such a triplet occur for *all* sufficiently large $p$? Denote by $\Omega(a, b)$ the least number such that a triplet is assured with $r \le \Omega(a, b)$ for all $p > p(a, b)$, and write $\Omega(a, b) = \infty$ if there is no finite number. For example, Emma Lehmer showed that $\Omega(1, 2) = \infty$, and more generally that $\Omega(a, b) = \infty$ if $(a, b) \equiv (1, 2) \pmod 3$ or if $(a, b) \equiv (1, 3), (2, 3)$, or $(2, 4) \pmod 5$ or if $(a, b) \equiv (1, 5), (2, 3)$, or $(4, 6) \pmod 7$. Is $\Omega(a, b)$ finite in all other cases? Emma Lehmer conjectures that it is finite if $a$ and $b$ are squares. Of course, $\Omega(a, b) = 1$ if $a$, $b$ are each one less than a square. As an example, let us see why $\Omega(5, 23) = 16$. If the triplets (1, 6, 24) and (4, 9, 27) are not all residues, then 6 and 3 are not, and 2 must be a residue. If the triplets (2, 7, 25) and (13, 18, 36) are not all residues, then 7 and 13 must be nonresidues. Under these circumstances, $(r, r + 5, r + 23)$ are not all residues for $1 \le r \le 15$, but when $r = 16$, (16, 21, 39) are residues.

Table 9 contains what are believed to be the (minimum) values of $\Omega(a, b)$. They provide good evidence for the conjectured finiteness in all cases except those already noted. Can an upper bound be obtained in terms of $a$ and $b$?

Table 9. Values of $\Omega(a, b)$ for $a < b \leq 25$.

| $b =$ / $a$ | 4 | 5 | 6 | 7 | 8 | 9 | 10 | 11 | 12 | 13 | 14 | 15 | 16 | 17 | 18 | 19 | 20 | 21 | 22 | 23 | 24 | 25 | $a$ |
|---|---|---|---|---|---|---|---|---|---|---|---|---|---|---|---|---|---|---|---|---|---|---|---|
| 1 | 45 | ∞ | 24 | 38 | ∞ | 84 | 26 | ∞ | ∞ | ∞ | ∞ | 77 | 35 | ∞ | ∞ | ∞ | ∞ | 15 | 35 | ∞ | 21 | 69 | 1 |
| 2 | ∞ | 25 | 20 | ∞ | ∞ | ∞ | ∞ | 70 | 30 | ∞ | ∞ | 54 | ∞ | ∞ | ∞ | ∞ | 25 | 98 | ∞ | ∞ | ∞ | ∞ | 2 |
| 3 | 174 | 39 | ∞ | ∞ | 1 | ∞ | 55 | ∞ | ∞ | 36 | 105 | 1 | ∞ | ∞ | 18 | 36 | 95 | ∞ | ∞ | ∞ | 1 | 51 | 3 |
| 4 |  | ∞ | ∞ | ∞ | ∞ | 91 | 36 | ∞ | ∞ | ∞ | ∞ | 126 | 60 | ∞ | 38 | 168 | ∞ | 90 | ∞ | ∞ | 66 | 77 | 4 |
| 5 |  |  | 49 | ∞ | ∞ | 121 | ∞ | 25 | 4 | ∞ | 28 | ∞ | ∞ | 64 | 110 | ∞ | 100 | 4 | ∞ | 16 | 64 | ∞ | 5 |
| 6 |  |  |  | 57 | ∞ | 33 | 30 | ∞ | 24 | ∞ | 42 | 60 | 36 | 38 | ∞ | 62 | 78 | 60 | 78 | ∞ | 45 | ∞ | 6 |
| 7 |  |  |  |  | ∞ | ∞ | 75 | ∞ | 74 | ∞ | ∞ | 27 | 9 | ∞ | ∞ | ∞ | ∞ | 70 | 42 | ∞ | ∞ | 45 | 7 |
| 8 |  |  |  |  |  | 66 | ∞ | ∞ | ∞ | ∞ | 30 | 1 | ∞ | ∞ | 77 | ∞ | 48 | ∞ | ∞ | 42 | 1 | ∞ | 8 |
| 9 |  |  |  |  |  |  | ∞ | 54 | ∞ | 42 | 66 | 57 | 66 | ∞ | 36 | 27 | 16 | 72 | ∞ | 21 | ∞ | 119 | 9 |
| 10 |  |  |  |  |  |  |  | ∞ | 60 | 85 | ∞ | 55 | ∞ | ∞ | 32 | 102 | ∞ | 77 | 26 | ∞ | 28 | 39 | 10 |
| 11 |  |  |  |  |  |  |  |  | 28 | ∞ | 119 | 49 | ∞ | 39 | ∞ | ∞ | ∞ | 64 | ∞ | ∞ | 25 | ∞ | 11 |
| 12 |  |  |  |  |  |  |  |  |  | ∞ | ∞ | ∞ | 65 | 98 | ∞ | ∞ | 36 | 4 | ∞ | ∞ | ∞ | 90 | 12 |
| 13 |  |  |  |  |  |  |  |  |  |  | ∞ | 42 | ∞ | ∞ | ∞ | 36 | ∞ | ∞ | ∞ | ∞ | 36 | ∞ | 13 |
| 14 |  |  |  |  |  |  |  |  |  |  |  | 35 | ∞ | ∞ | 42 | ∞ | 52 | 56 | ∞ | 64 | 81 | ∞ | 14 |
| 15 |  |  |  |  |  |  |  |  |  |  |  |  | 66 | 27 | 69 | ∞ | 49 | 25 | 99 | 110 | 1 | 105 | 15 |
| 16 |  |  |  |  |  |  |  |  |  |  |  |  |  | ∞ | ∞ | 102 | ∞ | 169 | 95 | ∞ | ∞ | 56 | 16 |
| 17 |  |  |  |  |  |  |  |  |  |  |  |  |  |  | ∞ | ∞ | 76 | 64 | ∞ | ∞ | ∞ | ∞ | 17 |
| 18 |  |  |  |  |  |  |  |  |  |  |  |  |  |  |  | 81 | ∞ | ∞ | ∞ | 40 | 185 | 144 | 18 |
| 19 |  |  |  |  |  |  |  |  |  |  |  |  |  |  |  |  | ∞ | 33 | ∞ | ∞ | 36 | 96 | 19 |
| 20 |  |  |  |  |  |  |  |  |  |  |  |  |  |  |  |  |  | 74 | ∞ | 40 | 25 | ∞ | 20 |
| 21 |  |  |  |  |  |  |  |  |  |  |  |  |  |  |  |  |  |  | 93 | ∞ | 70 | 100 | 21 |
| 22 |  |  |  |  |  |  |  |  |  |  |  |  |  |  |  |  |  |  |  | ∞ | ∞ | 98 | 22 |
| 23 |  |  |  |  |  |  |  |  |  |  |  |  |  |  |  |  |  |  |  |  | ∞ | ∞ | 23 |
| 24 |  |  |  |  |  |  |  |  |  |  |  |  |  |  |  |  |  |  |  |  |  | 63 | 24 |

What about patterns of four residues, $r$, $r+a$, $r+b$, $r+c$? Of course these won't necessarily occur if any of the four subpatterns of three residues aren't forced to do so. We need examine only $(a,b,c) = (2,5,6)$, $(1,6,7)$, $(1,4,9)$, $(5,6,9)$, $(1,6,10)$, $(1,7,10)$, ... where $\Omega(a,b)$, $\Omega(a,c)$, $\Omega(b,c)$, and $\Omega(b-a, c-a)$ are each known to be finite. Some corresponding values of $\Omega(a,b,c)$ are $\Omega(1,4,9) = 324$, $\Omega(1,4,15) = 675$, and, of course, $\Omega(3,8,15) = 1$.

But although $\Omega(1,6) = 24$, $\Omega(1,7) = 38$, $\Omega(5,6) = 49$, and $\Omega(6,7) = 57$, it appears that $\Omega(1,6,7) = \infty$. In fact the pattern $r$, $r+a$, $r+b$, $r+c$, $r+d$, with $(a,b,c,d) = (1,6,7,10)$ is such that, for each of the five subpatterns of four, $\Omega(1,6,7) = \Omega(1,6,10) = \Omega(1,7,10) = \Omega(5,6,9) = \Omega(6,7,10) = \infty$.

D. H. Lehmer and Emma Lehmer, On runs of residues, *Proc. Amer. Math. Soc.* **13** (1962) 102–106; *MR* **25** #2035.

D. H. Lehmer, Emma Lehmer and W. H. Mills, Pairs of consecutive power residues, *Canad. J. Math.* **15** (1963) 172–177; *MR* **26** #3660.

**F7.** Hugh Williams observes that if $p \equiv 3 \pmod 4$, then the equation $x^2 - py^2 = 2$ has integer solutions just if the congruence $w^2 \equiv 2 \pmod p$ does, i.e., just if $(\frac{2}{p}) = 1$, and asks for a cubic analog in the cases where $p \not\equiv \pm 1 \pmod 9$: $x^3 + py^3 + p^2z^3 - 3pxyz = 3$ is solvable just if $w^3 \equiv 3 \pmod p$ is. Barrucand and Cohn have shown that this is true for $p \equiv 2$ or $5 \pmod 9$. What about $p \equiv 4$ or $7 \pmod 9$? This is a special case of a more general conjecture of Barrucand. If true, it would be useful in abbreviating the calculations needed to find the fundamental unit (regulator) of the cubic field $\mathbb{Q}(\sqrt[3]{p})$.

P.-A. Barrucand and Harvey Cohn, A rational genus, class number divisibility and unit theory for pure cubic fields, *J. Number Theory* **2** (1970) 7–21.

H. C. Williams, Improving the speed of calculating the regulator of certain pure cubic fields, *Math. Comput.* **35** (1980) 1423–1434.

**F8.** Gary Ebert asks us to find the largest collection of quadratic residues $r_i \pmod{p^n}$ where $p^n \equiv 1 \pmod 4$ such that $r_i - r_j$ is a quadratic residue for all pairs $(i,j)$.

**F9.** A **primitive root**, $g$, of a prime $p$ is a number such that the residue classes of $g, g^2, \ldots, g^{p-1} = 1$ are all distinct. For example 5 is a primitive root of 23 because

$$5,\ 5^2 \equiv 2,\ 5^3 \equiv 10, \quad 4,\ -3, \quad 8,\ -6,\ -7, \quad 11, \quad 9,\ -1,$$
$$-5, \quad -2, \quad -10,\ -4, \quad 3,\ -8, \quad 6, \quad 7,\ -11,\ -9, \quad 1$$

all belong to different residue classes (mod 23).

Erdös asks: if $p$ is large enough, is there always a prime $q < p$ so that $q$ is a primitive root of $p$?

Basil Gordon asks if every odd prime $p$ has a primitive root $g > (p-1)/2$ which is itself a prime.

Given a prime $p > 3$, Brizolis asks if there is always a primitive root, $g$, of $p$ and $x$ $(0 < x < p)$ such that $x \equiv g^x \pmod{p}$. If so, can $g$ also be chosen so that $0 < g < p$ and $(g, p-1) = 1$?

Vegh asks whether, for all primes $p > 61$, every integer can be expressed as the difference of two primitive roots of $p$.

If $p$ and $q = 4p^2 + 1$ are both primes, Gloria Gagola asks if 3 is a primitive root of $q$ for all $p > 3$; is $p = 193$ the only odd prime for which 2 is not a primitive root of $q$; is $p = 653$ the only prime for which 5 fails to be both a quadratic residue and a primitive root of $q$; and is there a number, perhaps a function of $p$ (such as $2p - 1$ for large $p$) which is always a primitive root of $q$?

**F10.** Graham asks about the residue of $2^n \pmod{n}$. There are no solutions of $2^n \equiv 1 \pmod{n}$ with $n > 1$. $2^n \equiv 2 \pmod{n}$ whenever $n$ is a pseudoprime base 2 (see A12); in particular whenever $n$ is prime. The Lehmers have shown that the smallest solution of $2^n \equiv 3 \pmod{n}$ is $n = 4700063497 = 19 \times 47 \times 5263229$. Of course, $n$ has to be composite, and it is not divisible by any of 2, 3, 7, 17, 31, 41, 43, 73, . . . . What are these primes?

**F11.** What is the distribution of $1!, 2!, 3!, \ldots, (p-1)!, p!$ modulo $p$? About $p/e$ of the residue classes are not represented. Here are the missing ones for the first few values of $p$

$p = 2$ or 3, none. $\quad p = 5, \{-2\}. \quad p = 7, \{-2, 4\}.$
$p = 11, \{-2, \pm 3, \pm 4\}. \quad p = 13, \{-3, 4, -5\}.$
$p = 17, \{4, 5, -6, -7, -8\}. \quad p = 19, \{3, -5, -6, \pm 7, \pm 8\}.$
$p = 23, \{-3, -4, -6, -7, -8, 10\}.$
$p = 29, \{-2, -4, 7, -8, -9, -10, -11, -12, 13, -14\}.$
$p = 31, \{3, 4, 8, \pm 10, 11, 12, 13, 14\}.$
$p = 37, \{3, 4, \pm 5, -9, 10, 11, -14, \pm 15, -18\}.$

Until we reach the last two entries, we might be tempted to conjecture that there were always at least as many negative entries missing as positive ones. Are there infinitely many examples of each case? The value $p = 23$ is remarkable in that the only duplicates are $\pm 1$.

**F12.** For each $x$ $(0 < x < p)$ where $p$ is an odd prime, define $\bar{x}$ by $x\bar{x} \equiv 1 \pmod{p}$ and $0 < \bar{x} < p$. Let $N_p$ be the number of cases in which $x$ and $\bar{x}$ are of opposite parity. E.g., for $p = 13$, $(x, \bar{x}) = (1, 1)$, $\underline{(2, 7)}$, $(3, 9)$, $(4, 10)$, $\underline{(5, 8)}$, $\underline{(6, 11)}$, $(12, 12)$ so $N_{13} = 6$. D. H. Lehmer asks us to find $N_p$ or at least to say something nontrivial about it. $N_p \equiv 2$ or $0 \pmod{4}$ according as

$p \equiv \pm 1 \pmod 4$. $N_3 = N_7 = 0$; $N_5 = 2$; $N_{11} = N_{19} = N_{31} = 4$; $N_{13} = 6$; $N_{17} = 10$; $N_{23} = 12$, and $N_{29} = 18$.

**F13.** A system of congruences $a_i \pmod{n_i}$ $(1 \le i \le k)$ is called a **covering system** if every integer $y$ satisfies $y \equiv a_i \pmod{n_i}$ for at least one value of $i$. For example 0 (mod 2), 0 (mod 3), 1 (mod 4), 5 (mod 6), 7 (mod 12). If $c = n_1 < n_2 < \cdots < n_k$, Erdös offers \$500.00 for a proof or disproof of the existence of covering systems with $c$ arbitrarily large. Davenport and Erdös, and Fried found systems with $c = 3$; Swift with $c = 6$; Selfridge with $c = 8$; Churchhouse with $c = 10$; Selfridge with $c = 14$; Krukenberg with $c = 18$; and Choi with $c = 20$.

Erdös offers \$25.00 for a proof of the nonexistence of covering systems with all moduli $n_i$ odd, distinct, and greater than one; while Selfridge offers \$500.00 for an explicit example of such a system. More generally, "odd" can be replaced by "not divisible by the first $r$ primes." Jim Jordan offers comparable rewards to those mentioned above for solutions to the analogous problems for Gaussian integers (A15).

Erdös noted that you can have a covering system with all moduli $n_i$ distinct, squarefree, and greater than one by using the proper divisors of 210:

| $a_i$ | 0 | 0 | 0 | 1 | 0 | 1 | 1 | 2 | 2 | 23 | 4 | 5 | 59 | 104 |
|---|---|---|---|---|---|---|---|---|---|---|---|---|---|---|
| $n_i$ | 2 | 3 | 5 | 6 | 7 | 10 | 14 | 15 | 21 | 30 | 35 | 42 | 70 | 105 |

Krukenberg used 2 and squarefree numbers greater than 3. Selfridge asks if you can have such a system with $c \ge 3$ in place of $c = 2$. He observes that the $n_i$ cannot all be squarefree with at most two prime factors, but the above example shows that you do not need more than three.

It is easy, but not trivial, to prove that, for a covering system with distinct moduli, $\sum_{i=1}^{k} 1/n_i > 1$. The sum can be arbitrarily close to 1 if $n_1 = 3$ or 4. Selfridge and Erdös conjecture that $\sum 1/n_i > 1 + c_{n_1}$ where $c_{n_1} \to \infty$ with $n_1$.

Erdös also formulates the following conjecture. Consider all the arithmetic progressions of odd numbers, no term of which is of the form $2^k + p$. Is it true that all these progressions can be obtained from covering congruences? Are there infinitely many integers, not of the form $2^k + p$, which are not in such progressions? Which of

$$127, 149, 251, 331, 337, 373, 509, 599, 701,$$
$$757, 809, 877, 905, 907, 959, 977, 997, \ldots$$

are not in such?

S. L. G. Choi, Covering the set of integers by congruence classes of distinct moduli, *Math. Comput.* **25** (1971) 885–895; *MR* **45** #6744.

R. F. Churchhouse, Covering sets and systems of congruences, in *Computers in Mathematical Research*, North-Holland, 1968, 20–36; *MR* **39** #1399.

Fred Cohen and J. L. Selfridge, Not every number is the sum or difference of two prime powers, *Math. Comput.* **29** (1975) 79–81.

P. Erdös, Some problems in number theory, in *Computers in Number Theory*, Academic Press, 1971, 405–414; esp. pp. 408–409.
J. Haight, Covering systems of congruences, a negative result, *Mathematika* **26** (1979) 53–61; *MR* **81e**:10003.
J. H. Jordan, Covering classes of residues, *Canad. J. Math.* **19** (1967) 514–519; *MR* **35** #1538.
J. H. Jordan, A covering class of residues with odd moduli, *Acta Arith.* **13** (1967–68) 335–338; *MR* **36** #3709.
C. E. Krukenberg, PhD thesis, Univ. of Illinois, 1971, 38–77.
A. Schinzel, Reducibility of polynomials and covering systems of congruences, *Acta Arith.* **13** (1967) 91–101; *MR* **36** #2596.

**F14.** If a system of congruences is both covering and disjoint (each integer covered by just one congruence) it is called an **exact covering system**. Necessary, but not sufficient, conditions for a system to be exact are $\sum_{i=1}^{k} 1/n_i = 1$ and $(n_1, n_2, \ldots, n_k) > 1$ where the notation is as in F13. Znám confirmed a conjecture of Mycielski by further proving that if the l.c.m. $[n_1, n_2, \ldots, n_k] = \prod_j p_j^{\alpha_j}$, then $k \geq 1 + \sum \alpha_j(p_j - 1)$. He later generalized a result of Davenport, Mirsky, Newman, and Rado by proving that if $p$ is the least prime divisor of $n_k$, then $n_k = n_{k-1} = \cdots = n_{k-p+1}$. He conjectured that if there exist only *pairs* of equal moduli, then the moduli are all of the form $2^\alpha 3^\beta$, but later he and Schönheim gave counter-examples, such as

| | | | | | | | | |
|---|---|---|---|---|---|---|---|---|
| | 0, 1 | 2, 7 | 3, 8 | 13, 28 | 4, 9 | 14, 34 | 19, 39 | 59, 119 |
| mod | 5 | 10 | 15 | 30 | 20 | 40 | 60 | 120 |

Stein proved that if there is a single pair of equal moduli, the rest being distinct, then $n_i = 2^i$ $(1 \leq i \leq k-1)$, $n_k = 2^{k-1}$. Znám proved analogously that if there is a triple of equal moduli, the rest being distinct, then $n_i = 2^i$ $(1 \leq i \leq k-3)$, $n_{k-2} = n_{k-1} = n_k = 3 \times 2^{k-3}$. The main outstanding problem is to characterize exact covering congruences.

Porubsky asked if there is an "exactly $m$ times covering system" which is *not* the union of $m$ exact covering systems.

There are connexions with pseudoperfect numbers (B2) and Egyptian fractions (D11).

N. Burshtein and J. Schönheim, On exactly covering systems of congruences having moduli occurring at most twice, *Czechoslovak Math. J.* **24** (**99**) (1974) 369–372; *MR* **50** #4521.
A. S. Fraenkel, A characterization of exactly covering congruences, *Discrete Math.* **4** (1973) 359–366.
Bretislav Novák and Štefan Znám, Disjoint covering systems, *Amer. Math. Monthly* **81** (1974) 42–45.
Š. Znám, On Mycielski's problem on systems of arithmetical progressions, *Colloq. Math.* **15** (1966) 201–204; *MR* **34** #134.
Š. Znám, On exactly covering systems of arithmetic sequences, *Math. Ann.* **180** (1969) 227–232; *MR* **39** #4087.
Š. Znám, A simple characterization of disjoint covering systems, *Discrete Math.* **12** (1975) 89–91.

**F15.** Graham offers \$25.00 for settling the question: does $0 < a_1 < a_2 < \cdots < a_n$ imply that $\max_{i,j} a_i/(a_i, a_j) \geq n$?

Marica and Schönheim showed that if not, then not all the $a_i$ are squarefree. Winterle showed that $a_1$ is not prime, and Szemerédi that $n$ is not prime. Weinstein verified the statement if one of the $a_i$ is prime and not the arithmetic mean of two others, and Chein managed to dispense with this last condition. Vélez showed that $n - 1$ is not prime and that if a prime $p$ divides some $a_i$, then $p \leq n$. Boyle improved this to $p \leq (n-1)/2$ and to $p \leq (n-1)/3$ in many cases and produced some more complicated results.

R. D. Boyle, On a problem of R. L. Graham, *Acta Arith.* **34** (1978) 163–177.
E. Z. Chein, On a conjecture of Graham concerning a sequence of integers, *Canad. Math. Bull.* **21** (1978) 285–287; *MR* **80d**: 10024; *Zbl.* 392.10002.
P. Erdös, Problems and results in combinatorial number theory, in *A Survey of Combinatorial Theory*, North-Holland, 1973, 117–138.
R. L. Graham, Advanced problem 5749*, *Amer. Math. Monthly* **77** (1970) 775.
J. Marica and J. Schönheim, Differences of sets and a problem of Graham, *Canad. Math. Bull.* **12** (1969) 635–637.
R. J. Simpson, On a conjecture of R. L. Graham, *Acta Arith.* (to appear)
William Yslas Vélez, Some remarks on a number theoretic problem of Graham, *Acta Arith.* **32** (1977) 233–238.
Gerald Weinstein, On a conjecture of Graham concerning greatest common divisors, *Proc. Amer. Math. Soc.* **63** (1977) 33–38; *Zbl.* 369.10003.
Riko Winterle, A problem of R. L. Graham in combinatorial number theory, *Congressus Numerantium* I, Proc. Conf. Combin. Baton Rouge, Utilitas Math. Pub. 1970, 357–361; *MR* 42 #3051.

**F16.** Erdös defines $A(n, k)$ as $\prod p^a$ where the product is taken over primes $p$ less than $k$ and $p^a \| n$, and asks: is

$$\max_n \min_{1 \leq i \leq k} A(n+i, k) = o(k) \qquad ?$$

He remarks that it is easy to show that it is $O(k)$. Is

$$\min_n \max_{1 \leq i \leq k} A(n+i, k) > k^c$$

for every $c$ and sufficiently large $k$? Is

$$\sum_{i=1}^{k} 1/A(n+i, k) > c \ln k \qquad ?$$

**F17.** Alf van der Poorten asks for a proof that

$$\zeta(4) \left[ = \sum_{n=1}^{\infty} \frac{1}{n^4} = \frac{\pi^4}{90} \right] = \frac{36}{17} \sum_{n=1}^{\infty} \frac{1}{n^4 \binom{2n}{n}}.$$

It is known that

$$\sum_{n-1}^{\infty} \frac{1}{\binom{2n}{n}} = \frac{2\pi\sqrt{3}+9}{27}, \qquad \sum_{n-1}^{\infty} \frac{1}{n\binom{2n}{n}} = \frac{\pi\sqrt{3}}{9},$$

$$\zeta(2)\left[= \sum_{n=1}^{\infty} \frac{1}{n^2} = \frac{\pi^2}{6}\right] = 3\sum_{n=1}^{\infty} \frac{1}{n^2\binom{2n}{n}},$$

$$2(\sin^{-1} x)^2 = \sum_{n=1}^{\infty} \frac{(2x)^{2n}}{n^2\binom{2n}{n}} \quad \text{and} \quad \zeta(3)\left[= \sum_{n=1}^{\infty} \frac{1}{n^3}\right] = \frac{5}{2}\sum_{n=1}^{\infty} \frac{(-1)^{n-1}}{n^3\binom{2n}{n}},$$

but how does one prove the $\zeta(4)$ result? It has been noticed numerically and is mentioned by Comtet.

Louis Comtet, *Advanced Combinatorics*, Dreidel, Dordrecht, 1974, p. 89.
Alfred van der Poorten, A proof that Euler missed . . . Apery's proof of the irrationality of $\zeta(3)$. An informal report, *Math. Intelligencer* **1** (1979) 195–203.
Alfred J. van der Poorten, Some wonderful formulas . . .an introduction to polylogarithms, *Proc. Number Theory Conf.*, Queen's Univ., Kingston, 1979, 269–286; *MR* **80i**:10054.

**F18.** If $a_1, a_2, \ldots, a_n$ are $n$ numbers (of any sort), how big is the set of their sums and products in pairs?

$$¿ \qquad |\{a_i + a_j\} \cup \{a_i a_j\}| > n^{2-\varepsilon} \qquad ?$$

Erdös and Szemerédi have proved that the cardinality is $> n^{1+\varepsilon}$.

P. Erdös, Some recent problems and results in graph theory, combinatorics and number theory, *Congressus Numerantium XVII*, Proc. 7th S.E. Conf. Combin. Graph Theory, Comput. Boca Raton, 1976, 3–14 (esp. p. 11).

**F19.** If $n$ is sufficiently large and is written in the form $n = a + b + c$, $0 < a < b < c$ in every possible way, are all the products $abc$ distinct?

J. Riddell and H. Taylor asked if, among the partitions of $n$ into *distinct primes*, the one having the maximum *product* of parts is necessarily one of those with the maximum *number* of parts, but Selfridge gave a negative answer. For example

$$\begin{aligned} 319 &= 2 + 3 + 5 + 7 + 11 + 13 + 17 + 23 + 29 + 31 + 37 + 41 + 47 + 53 \\ &= 3 + 5 + 11 + 13 + 17 + 19 + 23 + 29 + 31 + 37 + 41 + 43 + 47 \end{aligned}$$

but the partition with the smaller number of parts gives the largest possible product. Is this the least counter-example? Can the cardinalities of the two sets differ by an arbitrarily large amount?

**F20.** A number $x$ may be expressed as a **continued fraction**

$$x = a_0 + \cfrac{b_1}{a_1 + \cfrac{b_2}{a_2 + \cfrac{b_3}{a_3 + \cdots}}}$$

which, out of kindness to the typesetter, is usually written

$$x = a_0 + \frac{b_1}{a_1 +}\,\frac{b_2}{a_2 +}\,\frac{b_3}{a_3 +}\cdots.$$

When the numerators $b_i$ are all 1, the continued fraction is called **simple**. It may be finite or infinite, but if $x$ is rational it is finite. In this case there are two possible forms, one of which has its last **partial quotient**, $a_k$, equal to 1:

$$\frac{7}{16} = \frac{1}{2+}\,\frac{1}{3+}\,\frac{1}{2} = \frac{1}{2+}\,\frac{1}{3+}\,\frac{1}{1+}\,\frac{1}{1}.$$

Not every number $n$ can be expressed as a sum of two positive integers $n = a + b$ so that the continued fraction for $b/a$ has all its partial quotients equal to 1 or 2. For 11, 17, and 19 we can have

$$\frac{4}{7} = \frac{1}{1+}\,\frac{1}{1+}\,\frac{1}{2+}\,\frac{1}{1}, \quad \frac{5}{12} = \frac{1}{2+}\,\frac{1}{2+}\,\frac{1}{2}, \quad \text{and} \quad \frac{7}{12} = \frac{1}{1+}\,\frac{1}{1+}\,\frac{1}{2+}\,\frac{1}{2}$$

but 23 can't be so expressed. However Leo Moser conjectured that there is a constant $c$ such that every $n$ can be so expressed, with the sum of the partial quotients, $\sum a_i$, less than $c \ln n$.

**F21.** Bohuslav Divis asks for a proof that in each real quadratic field there is an irrational number whose simple continued fraction expansion has all its partial quotients 1 or 2. He also asks the same question with 1 and 2 replaced by any pair of distinct natural numbers.

**F22.** Is there an algebraic number (of degree greater than two) whose continued fraction has unbounded partial quotients? Does *every* real algebraic number of degree greater than two have unbounded partial quotients? Ulam asks particularly about the number $\xi = 1/(\xi + y)$ where $y = 1/(1 + y)$.

Littlewood observes that if $\theta$ has a continued fraction with bounded partial quotients $a_n$, then $\lim\inf n|\sin n\theta| \le A(\theta)$, where $A(\theta)$ is not zero (though it is for almost all $\theta$). He asks if

$$\lim\inf n|\sin n\theta \sin n\phi| = 0$$

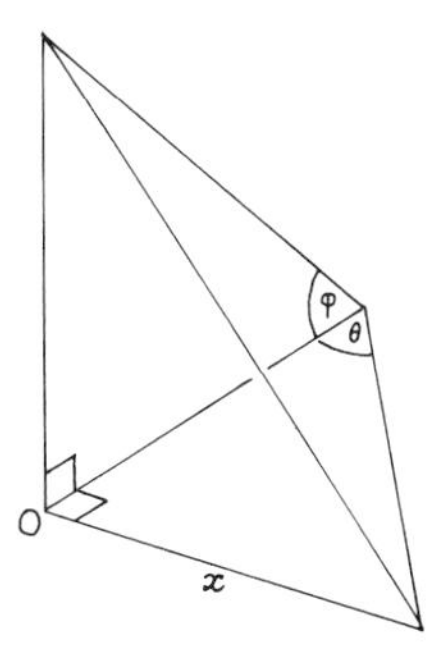


Figure 17. Rectangular Tetrahedron.

for all real $\theta$ and $\phi$? It is for almost all $\theta$ and $\phi$. On the other hand

$$\liminf n^{1+\varepsilon}|\sin n\theta \sin n\phi| = \infty$$

for almost all $\theta$ and $\phi$. Cassels and Swinnerton-Dyer treat a dual problem and show incidentally that $\theta = 2^{1/3}$, $\phi = 4^{1/3}$ does *not* give a counter-example. Davenport suggests that a computer might help with proving that

$$|(x\theta - y)(x\phi - z)| < \varepsilon$$

has solutions for *every* $\theta$, $\phi$ when, for example, $\varepsilon = \frac{1}{10}$ or $\frac{1}{50}$.

A connected problem concerns the number $N(x)$ of lattice points in a tetrahedron (Figure 17) which is rectangular at $O$ with $x$ large. If the volume $V(x) = N(x) - E(x)$, then the error $E(x)$ is $O((\ln x))^k$ with $k$ something like 2 or 3, for almost all $\theta$, $\phi$, but nothing better than $O(x \ln x)$ is known for any particular pair $\theta$, $\phi$.

J. W. S. Cassels and H. P. F. Swinnerton-Dyer, On the product of three homogeneous linear forms and indefinite ternary quadratic forms, *Philos. Trans. Roy. Soc. London Ser. A* **248** (1955) 73–96; *MR* **17**–14.

Harold Davenport, Note on irregularities of distribution, *Mathematika* **3** (1956) 131–135; *MR* **19**, 19.

John E. Littlewood, *Some Problems in Real and Complex Analysis*, Heath, Lexington Mass., 1968, 19–20, Problems 5,6.

**F23.** Problem 1 of Littlewood's book asks how small $3^n - 2^m$ can be in comparison with $2^m$. He gives as an example

$$\frac{3^{12}}{2^{19}} = 1 + \frac{7183}{524288} \approx 1 + \frac{1}{73}$$

(the ratio of D$^\#$ to E$^\flat$).

Croft asks the corresponding question for $n! - 2^m$. The first few best approximations to $n!$ by powers of 2 are

| $5!$ | $20!$ | $22!$ | $24!$ | $61!$ | $63!$ | $90!$ |
|---|---|---|---|---|---|---|
| $2^7$ | $2^{61}$ | $2^{70}$ | $2^{79}$ | $2^{278}$ | $2^{290}$ | $2^{459}$ |
| $-1.34$ | $+1.26$ | $-0.10$ | $+0.046$ | $+0.023$ | $-0.0017$ | $-0.0007$ |

where the third row is the percentage error in the exponent. Erdös observes that $n! = 2^a \pm 2^b$ only when $n = 1, 2, 3, 4$ and 5.

**F24.** Sin Hitotumatu asks for a proof or disproof that, apart from $10^{2n}$, $4 \times 10^{2n}$, and $9 \times 10^{2n}$, there are only finitely many squares with just two different decimal digits, such as $1444 = 38^2$, $7744 = 88^2$, $11881 = 109^2$, $29929 = 173^2$, $44944 = 212^2$, $55225 = 235^2$, and $9696996 = 3114^2$.

**F25.** In the sequence 679, 378, 168, 48, 32, 6, each term is the product of the decimal digits of the previous one. Neil Sloane defines the **persistence** of a number as the number of steps (5 in the example) before the number collapses to a single digit. The smallest numbers with persistence

1 2 3 4 5 6 7 8 9 10 11

are

10 25 39 77 679 6788 68889 2677889 26888999 3778888999 277777788888899

There is no number less than $10^{50}$ with persistence greater than 11. Sloane conjectures that there is a number $c$ such that no number has persistence greater than $c$.

In base 2 the maximum persistence is 1. In base 3 the second term is zero or a power of 2. It is conjectured that all powers of 2 greater than $2^{15}$ contain a zero when written in base 3. This is true up to $2^{500}$. The truth of this conjecture would imply that the maximum persistence in base 3 is 3.

Sloane's general conjecture is that there is a number $c(b)$ such that the persistence in base $b$ cannot exceed $c(b)$.

Erdös modifies the problem by letting $f(n)$ be the product of the *nonzero* decimal digits of $n$, and asks how fast one reaches a one-digit number, and for which numbers is the descent slowest. He says it is easy to prove that $f(n) < n^{1-c}$, so that at most $c \ln \ln n$ steps are needed.

N. J. A. Sloane, The persistence of a number, *J. Recreational Math.* **6** (1973) 97–98.

**F26.** Let $f(n)$ be the least number of ones that can be used to represent $n$ using ones and any number of + and × signs. For example,

$$80 = (1 + 1 + 1 + 1 + 1) \times (1 + 1 + 1 + 1) \times (1 + 1 + 1 + 1)$$

so $f(80) \leq 13$. It can be shown that $f(3^k) = 3k$ and $3 \log_3 n \leq f(n) < 5 \log_3 n$ where the logs are to base 3. Does $f(n) \sim 3 \log_3 n$?

J. H. Conway and M. J. T. Guy, $\pi$ in four 4's, *Eureka* **25** (1962) 18–19.
Popken and K. Mahler, On a maximum problem in arithmetic, *Nieuw Arch. voor Wisk.* (3) **1** (1953) 1–15.

**F27.** The **Farey series** of order $n$ consists of all positive rational numbers in their lowest terms, with numerators and denominators not exceeding $n$, arranged in order of magnitude. For example, the Farey series of order 5 is

$$\tfrac{1}{5}\ \tfrac{1}{4}\ \tfrac{1}{3}\ \tfrac{2}{5}\ \tfrac{1}{2}\ \tfrac{3}{5}\ \tfrac{2}{3}\ \tfrac{3}{4}\ \tfrac{4}{5}\ \tfrac{1}{1}\ \tfrac{5}{4}\ \tfrac{4}{3}\ \tfrac{3}{2}\ \tfrac{5}{3}\ \tfrac{2}{1}\ \tfrac{5}{2}\ \tfrac{3}{1}\ \tfrac{4}{1}\ \tfrac{5}{1}.$$

The determinant formed from the numerators and denominators of two adjacent fractions is $\pm 1$. Mahler regards the members of the sequence as positive real roots of linear equations whose coefficients have g.c.d. 1 and do not exceed $n$, and obtained the following apparent generalization to quadratic equations. List the coefficients $(a, b, c)$ of the quadratic equations

$$ax^2 + bx + c = 0,\ a \geq 0,\ (a, b, c) = 1,\ b^2 \geq 4ac,\ \max\{a, |b|, |c|\} \leq n,$$

which have positive real roots, in order of the size of the roots. Then the third-order determinant (see F28) formed from any three consecutive rows of $a$, $b$, $c$ is always 0 or $\pm 1$.

We have started Table 10, for $n = 2$, with the entries 0, 1, 0, corresponding to the root zero; adopted a suggestion of Selfridge of duplicating rational roots to avoid trivial exceptions; and ended with 0, 0, 1. The entry in any row in the last column is the value of the determinant formed from that row and the ones immediately above and below it.

Table 10. Determinants of Coefficients of Quadratic Equations Arranged in Order of Size of Roots.

| $a$ | $b$ | $c$ | root | determinant |
|---|---|---|---|---|
| 0 | 1 | 0 | 0 | |
| 2 | 2 | −1 | $(\sqrt{3}-1)/2$ | 1 |
| 1 | 2 | −1 | $\sqrt{2}-1$ | 0 |
| 0 | 2 | −1 | $\frac{1}{2}$ | 0 |
| 0 | 2 | −1 | $\frac{1}{2}$ | 0 |
| 1 | 1 | −1 | $(\sqrt{5}-1)/2$ | 0 |
| 2 | 0 | −1 | $\sqrt{2}/2$ | −1 |
| 1 | 2 | −2 | $\sqrt{3}-1$ | −1 |
| 2 | 1 | −2 | $(\sqrt{17}-1)/4$ | 1 |
| 0 | 1 | −1 | 1 | 0 |
| 0 | 1 | −1 | 1 | 0 |
| 2 | −1 | −2 | $(\sqrt{17}+1)/4$ | 0 |
| 2 | −2 | −1 | $(\sqrt{3}+1)/2$ | 1 |
| 1 | 0 | −2 | $\sqrt{2}$ | −1 |
| 1 | −1 | −1 | $(\sqrt{5}+1)/2$ | 1 |
| 0 | 1 | −2 | 2 | 0 |
| 0 | 1 | −2 | 2 | 0 |
| 1 | −2 | −1 | $\sqrt{2}+1$ | 1 |
| 1 | −2 | −2 | $\sqrt{3}+1$ | 0 |
| 0 | 0 | 1 | $\infty$ | |

Can the result be proved? Does it generalize to fourth-order determinants associated with cubic equations?

## F28. The **third-order determinant**

$$\begin{vmatrix} a_1 & a_2 & a_3 \\ a_4 & a_5 & a_6 \\ a_7 & a_8 & a_9 \end{vmatrix}$$

may be defined as $a_1(a_5a_9 - a_6a_8) - a_2(a_4a_9 - a_6a_7) + a_3(a_4a_8 - a_5a_7)$.

Basil Gordon asks for whole numbers $a_1, a_2, \ldots, a_9$, none of them 0 or $\pm 1$ so that

$$\begin{vmatrix} a_1 & a_2 & a_3 \\ a_4 & a_5 & a_6 \\ a_7 & a_8 & a_9 \end{vmatrix} = 1 = \begin{vmatrix} a_1^2 & a_2^2 & a_3^2 \\ a_4^2 & a_5^2 & a_6^2 \\ a_7^2 & a_8^2 & a_9^2 \end{vmatrix}.$$

## F29. Given a prime $p$, find pairs of functions $f(x)$, $g(x)$ such that one of the congruences $f(x) \equiv n$, $g(x) \equiv n \pmod p$ is solvable for all integers $n$. A trivial example is $f(x) = x^2$, $g(x) = ax^2$, where $a$ is a quadratic nonresidue of the odd prime $p$. Mordell gives the further example $f(x) = 2x + dx^4$, $g(x) = x - 1/4dx^2$, where $d$ is any integer prime to $p$ and $1/z$ is defined as $\bar{z}$, where $z\bar{z} \equiv 1 \pmod p$.

## F30. It was noted in D1 that no nontrivial solution of $a^5 + b^5 = c^5 + d^5$ is known. In fact $x^5$ is almost certainly an answer to the following unsolved problem of Erdös. Find a polynomial $P(x)$ such that all the sums $P(a) + P(b)$ $(a < b)$ are distinct.

# Index of Authors Cited

The names appearing here are those of authors whose works are referred to in this volume. References occur at the end of each problem (e.g. D11, pp. 90–93); in the Introduction, I (pp. 1–2) and at the beginning of Sections A (pp. 3–4), D (p. 79), E (p. 110) and F (p. 132). Mentions unsupported by references are listed in the General Index.

# General Index

Names appear here if their mention in this volume is unsupported by references. Single letter entries refer to the Introduction, I(pp. 1–2) and to the beginning of Sections A (pp. 3–4), E (p. 110) and F (p. 132).